Act Now, Apologize Later

ACT NOW, APOLOGIZE LATER

Adam Werbach

Cliff Street Books

An Imprint of HarperCollins*Publishers*

HarperCollins books may be purchased for educational, business, or sales promotional use. For information please write: Special Markets Department,
HarperCollins Publishers, Inc., 10 East 53rd Street,
New York, NY 10022.

FIRST EDITION

Designed by Interrobang Design Studio

Library of Congress Cataloging-in-Publication Data

Werbach, Adam.
    Act now, apologize later / Adam Werbach.
    — 1st ed.
        p.   cm.
    ISBN 0-06-017550-8
    1. Werbach, Adam.   2. Environmentalism—
United States.   3. United States—Environmental
conditions.   4. Environmental responsibility—
United States.   5. Environmentalists—United
States—Biography.   I. Title.
GE56.W47A3   1997
363.7—dc21   97-39539

97 98 99 00 01 ❖/HC 10 9 8 7 6 5 4 3 2

Contents

Acknowledgments

This book is in your hands, thanks to the help of Aaron Presbrey, whose extraordinary writing skills, editing ability, humor, perseverance, and stamina brought this project to fruition.

This book is dedicated to Caleb, Trish, and Dylan Rick.

My thanks go to the volunteer leaders, staff, and members of the Sierra Club. It gives me great faith to know that right now in neighborhoods across North America you are fighting to protect our future. You are my heroes.

Dan "River-Rafting" Strone and his assistant, Deb Bard, guided me through the turbulent waters of the book business, which make the rapids of Cataract Canyon seem tame in comparison.

My extremely patient, direct, and kind editor, Diane Reverand, Meaghan Dowling, and David Flora from HarperCollins pulled together a rough-hewn manuscript, a cover from my favorite "agitational art" artist, and paper containing Asian bamboo in just a few weeks.

My editing posse stopped their lives for a few harried days and gave it their best. Dad and Mom, my brother, Kevin, Janice Presbrey, Stasia "Merde" Blyskal, Jessica "Too-Fine" Tully, Mark "Raisin-Locks" Fraioli, David "Martini" Sabban, Tom "Grizzly" Elliott, Emily "hEMpILY" Miggins, Helen and Nate "Grandma and

Grandpa" Landgarten, Laura "NREPA!" Hoehn, Jess "Barbie" Levin, and Michelle "Die-Disney-Die" Sheaffer.

The following people helped provide support, information, and inspiration for the writing of this book: David Bower, Isaac Peace Hazard, Dave Foreman, Erin Potts, Adam "Ruby-Repairer" Yauch, Edith and Henry Everett, Stephanie Jowers, Amory Lovins, Vice President Gore, Alison Chase, Denny Shaffer, Randy Hayes, Paul Loeb, Aaron May, Harry and 'Becca Dalton, Woody Harrelson, Pierce Flynn, Huey Johnson, Holly Minch, Todd Hettenbach, Alvin Mars, Dr. Rich Ingebretsen, Jann Wenner, Bobby Weir, Colleen McCabe, Derek and Colleen Bellin, Robbie Cox, Daniel Stern, Susan Holmes, Bob Bingarnan, Linda Powell, Wendy Van Norden, my grandparents, and Derby the Bear.

The paper for this book contains some ingredients not made from trees. While bamboo paper isn't the best answer, it's a step away from destroying our forests for pulp. By buying this book you have supported alternative papermaking. Thank you.

Give all of the people named above the credit for the good things in the book. I'll take the blame.

To support the Sierra Club or for more information, contact:

Adam Werbach
c/o Sierra Club
Act Now, Apologize Later
85 2nd Street, 2nd Floor
San Francisco, CA 94105
www.sierraclub.org

Introduction

You can learn a lot about society from the way we treat our kids' TV characters.

Oscar the Grouch, a smelly green whiner who lives in a garbage can, has been the star of *Sesame Street* for years. Yogi Bear, the smiling optimist who believes there's a picnic basket around every corner, was canceled. It's not cool to be positive.

I don't understand our attraction to grouchy cartoon monsters and grumpy people. I'm sick of them complaining about traffic and clearcutting and education and poverty without lifting a Lee press-on nail to help. I know we have problems.

Wal-Mart trounces the local drugstore, churches and temples scramble to retain membership rather than healing society, tract housing muscles out our wilderness, and the Psychic Friends Network is America's newest best friend. Parents in Seattle rush their children to the hospital because of poisoned hamburgers. Grizzly bears in Montana number

fewer than one thousand. Elderly people in Chicago
die because of air choked with smog. Beijing and
Toronto offer oxygen bars where, for twenty dollars
an hour, you can breathe healthful air. Global warming melts the ice caps in the Arctic Circle down to the
blue ice, causing dust devils—blistering dirt torna-
does—to sweep across the tundra like a pack of cartoon Tasmanian Devils, destroying everything in
their path. The native Inuit don't have a word in
their language for dust devils.

It's almost as if the sky were falling. Welcome to a
new millennium.

I don't claim to know all the answers. I'm still
struggling to create a life where I can take care of
my friends and family, do my job well, and make
time to jam on my guitar. I have learned one thing:
Complaining gets me nowhere. I've decided to take
an active role in tackling the challenges that I face.
Last year, at age twenty-three, I was elected to the
office of president of the Sierra Club—America's oldest and largest grassroots environmental organization, with more than a half-million members, chapters in every state, and a history that began with the
protection of Yosemite National Park. I've chosen to
get involved.

I could give you lists of society's problems as long
as the three different J. Crew catalogs you receive
in the first two weeks of December. But when a
man named after a lizard, or a Newt anyway, tries

to destroy the Endangered Species Act, it doesn't take Starsky and Hutch to figure out that we've got a problem. My generation's defining song is "(You Gotta) Fight for Your Right (to Party!)" by the Beastie Boys, and even I know there's a problem. Complaining won't get us the world we want to live in.

I focus on the positive. That's why I believe in the environmental movement. In the last twenty-five years, the environmental movement has shown us that neighbors, working together, can find solutions to our challenges. I've experienced the changes; so have you.

Growing up in Los Angeles, I remember leaving the water of Santa Monica Bay with my feet tarred by oil spilled recklessly in the ocean. It took two weeks of Mr. Bubble and scrubbing to get it off. Oil spills off the coast of California were commonplace in the 1970s. Today, because of environmental legislation, you won't get "tarfooted" when you go to the beach. That's progress.

In our rush to complain, we often miss the hope that a success like the protection of the California coast can bring. Successes keep our faith in the future—the belief that tomorrow will be a better day than today—alive.

In order to reinstill America's faith in the future and save our environment we need to take five steps forward.

1. *Nature*

We need a dose of natural redemption. Weekend hikers and mountain bikers join millions of Americans in choosing wilderness to fulfill their spiritual and physical needs. Nature offers a nourishing sense of place, where you can reconnect to the wild within you. If you want to be told how to feel and what to do, go be a dittohead in Rush Limbaugh's studio audience. If you want to test your limits and grow, work to put yourself atop a waterfall, where a liberal arts degree is meaningless. A complete person leaves concrete and goes to a natural area at least twice a month. It can be an urban park or the wilds of Alaska.

2. *Youth*

Let's junk the Generation X myth and move on. For every slacker you show me, I'll tell you stories of young people like Rick Takeda, a surfer from Newport Beach who helped stop Newt's war on the environment by getting a million Americans to sign an environmental bill of rights. We can learn from the hope, wisdom, and daring of young people.

3. *Celebration*

Beth Gausman walked by a parked UPS truck and was crippled for life by toxic fumes that leaked from a crushed package. Instead of retreating, she pushed for legislation to prevent similar tragedies throughout America. She had every reason to give up on the world because of her bad luck. She chose to get active. If Beth could summon the courage to lead, so can you. I was once asked what the first thing would be that I would do if I were appointed secretary of the interior. I said that I'd throw the biggest party America had ever seen to celebrate our national parks. Unless we pause to celebrate the success of individuals like Beth, and the achievements of our nation, we'll lose the faith to carry on.

4. Communication

We need to move beyond the arrogance of environmentalism. Anyone who wants the basics, like breathing air that won't make you choke, is an environmentalist. That only leaves out Vegas lounge singers, and even they need to breathe sometimes. If everyone is an environmentalist, then environmental leaders need to speak in a language that everyone can understand. We need everyone, Republican and Democrat, liberal and conservative, young and old, to come together to fight for the planet that sustains us.

5. Community

Ever since the interstate highway system was established, our communities have been fragmenting and crumbling. Today, the roots of environmental problems stem from the disintegration of our communities. Megastores invade towns across America, leaving decaying communities in their wake. Will your last local farm give way to a strip mall? Will we preserve our cities and hometowns or lose them forever? It's time to form a new movement of radical locals to reclaim the character, economy, and wildness of America. It's time for environmentalists to care for the Bronx as much as they care for the Grand Canyon.

Some people will say that we shouldn't bother stewarding the environment now because the technology and wisdom of future generations will save them from the mess that we're creating. Unfortunately the problems are ours now. Global warming has begun. It's already accelerating the spread of malaria and dengue fever. Our beaches are beginning to disappear.

I can't justify leaving a mess like this for my kids to clean up.

I'm already scared enough about having kids. I'm scared about whether I can answer their questions about sexually transmitted diseases, racism, and the national debt. I'm scared that my kids will hate me for creating a world where their horizons are limited. There will be no baseball in the park if global warming continues. There will be no camping trips in Yosemite unless we protect it. Unless we act now, we will move into an age of regrets.

There is no time to lose. We've come to the point where we need new leaders, leaders who don't know all of the answers, leaders who will make honest mistakes. Unless we act now, the things we value will be destroyed. When you undertake a righteous challenge, it's easier to ask for forgiveness than permission.

Shortly after I was elected president of the Sierra Club, my phone woke me at 6:30 A.M. "Yeah?" I growled.

"Mr. Werbach, this is the White House; we'd like you to come to meet the president to discuss your vision."

"Yeah, right, who is this?" I responded, certain a friend was messing with me.

"Can you be here tomorrow?"

I traveled all night, sitting next to a child with the lungs of Pavarotti. By 9:15 the next morning I was in a blue suit, sitting in on a policy discussion at the White House. The vice president was asked a question about trading air-emission permits. I closed my eyes, just for a second. The next thing I knew I was asleep. Until . . .

Out of a haze, I heard the vice president say, "Adam, am I boring you?" Everyone laughed. That is, everyone laughed except me.

"No, I'm sorry," I said lamely to the veep. This was not going well.

The policy briefing ended and I was hustled into

an elegant room to await my audience with the president. Out of the corner of my eye, I noticed a pile of extravagant, chlorine-bleached, white paper napkins sporting the golden embossed Seal of the President of the United States. I reached over and started stealing them. They'd make great souvenirs, and with the message I was planning to give to the president this was about as close as I'd come to sleeping in the Lincoln bedroom.

Suddenly, a stiff hand clenched my shoulder. I spun around to see the smile of Vice President Al Gore. I froze, expecting him to say, "Hey, kid, first you fall asleep during my presentation, and then you come to my boss's house and start stealing napkins?"

"What took you so long to get here?" he said, laughing at the public's fixation on my age. He seemed loose and relaxed, but more tired than he looked when I had met him five years before.

While writing *Earth in the Balance,* he had come to Los Angeles to meet with a small group of people to talk about the future of the environmental movement. After the meeting he shook my hand and said, "I'll be keeping an eye on you, Adam," and then sprinted off to his next power talk.

Now we were in the White House together. Of course, he had become vice president and I still had to show three forms of ID and submit to an X-ray photograph of my notepad before they let me in. It was still nice to be there.

"Mr. Vice President, the better question is"—I smiled back at him—"what is taking *you* so long?"

He slapped me on the back and chuckled. I thought to myself, "I'm going to throw an outrageous party the night this guy takes over."

Americans are environmentally aware. Children easily rattle off the codes of recyclable plastic and tell you that it's not cool to have Flipper in their tuna. While the American educational system lags behind the rest of the world in math, American kids know more about ecology than their counterparts in Europe and Asia. But our environmental challenge requires more than awareness.

While well-intentioned, the book *50 Simple Things You Can Do to Save the Earth* knocked the environmental movement backwards. Of course there are fifty simple things you can do to help save the environment, but there are one hundred hard things, too. The book implies that the problems facing our world—loss of wilderness, unsafe tap water, ozone depletion—are the public's fault, and if we just take shorter showers everything will be okay. That's blaming the victim for the problem. We each have a role to play, but our challenge is to hold the big boys accountable. The U.S. government is the largest polluter in the world, not your parents because they don't separate brown and green glass. Shell Oil creates more problems than Harry's Happy Hamburgers ever will by using Styrofoam boxes to wrap its

HappyBurgers. It's important to take small steps in your own life, but not at the expense of the big ones that society must take as a whole.

It took a whole nation to save a burning river in Ohio. A few years before I was born, the Cuyahoga River caught on fire and burned for days. The slow-moving water shot searing flames and putrid smoke high into the air. The Cuyahoga had been used as a dumping ground for toxic sludge.

The burning river catalyzed American action. Everyone knows that water shouldn't catch fire. Now, the Cuyahoga is coming back. It supports a world of life. New businesses bud along the banks of the river next to water lilies and black-eyed Susans.

The Cuyahoga River was reclaimed thanks to the work of a community that wanted change. They demanded that the river be cleaned. In a world where people still buy cosmetic breast implants and the networks canceled Dennis Miller's talk show and Michael Moore's *TV Nation,* we have one less problem. Because of the Clean Water Act and the Safe Drinking Water Act, both inspired by the Cuyahoga, most Americans can turn on the tap and safely enjoy a glass of water. If we can make a river that once burned flow free again, we can bring hope back to our communities and our lives.

This book is not about all of the problems we face. That's what TV is for. This book is full of stories about ordinary people who have accomplished extraordinary successes because of their dedication to their communities and to the future. From rural priests to animal trackers, from a twelve-year-old girl in California to three elderly women in Georgia, from senators to surfers and from Woody Harrelson to llama riders, an incredible array of people give us a thousand reasons to be hopeful.

Pick up the book on a sunny afternoon and read it straight through to travel to our world's natural wonders, hear the wisdom of the world's most successful environmentalists, learn the secrets of successful leaders, and laugh as slimy politicians get what they deserve. Put it in the bathroom to catch a quick fable here and there or to jump for a few minutes into the heart of a breathtaking canyon. Fall asleep with it after a long day and pick it right back up at any page the next day. Pass it on to a friend as you add your own story, adventure, and success.

The world doesn't need everyone to be a professional environmentalist. It needs everyone to demand basic environmental rights—clean air, water, and food, and as much wilderness as your parents left you.

The environmental movement is one of our great civic success stories. Thousands of people have involved themselves in the movement to protect their neighborhoods and their world. It's time to take their example and to spread the word. We can't afford not to.

Act now, apologize later.

SECTION ◇ 1

Revenge of the Tarzanian

James Watt Made Me an Environmentalist

I learned to sign my name in the second grade. I already knew how to scratch out my name in printed letters, but I wanted to make a statement with my signature. It didn't matter whether you could distinguish it from the scratchings of a monkey on an Etch-A-Sketch. It was the best thing I owned, and I was going to do with it what I pleased. Trying to prove that I was stylin', I scrawled my name on every blank surface in school. My Trapper Keeper notebook, once jammed full of puffy plastic stickers, now sported the curling sweep of countless attempts to write my name. Every desk, door, and doorknob was fair game.

I remember returning home from a day of second grade to find the ultimate signature game, a petition sitting on my parents' coffee table. It said something about Watt on it. I knew what *Watt* meant. I watched *The Electric Company*.

"Finally," I thought, "I'll have something cool for show-and-tell!"

How was I to know that James Watt was the secretary of the interior? How was I to know that the interior even needed a secretary? I had no idea that the
interior included our national parks. I certainly
didn't know that James Watt wanted to develop the
Bob Marshall Wilderness Area. "The Bob" is 1.5 million acres of breathtaking land on the western side
of the Continental Divide in Montana. It is one of the
last refuges in the world for the grizzly bear. When
Watt was appointed Ronald Reagan's secretary of
the interior, he immediately offered oil and gas leases
for "The Bob."

Watt's idea of bettering the national park system
was to expand the services offered: more hot dog
stands, motels, parking lots, movie theaters, Ferris
wheels, ski slopes, and Pizza Huts. What's a national
park without pewter collector spoons?

It took the Sierra Club about three seconds to
decide that Watt had to go. Carl Pope, the Club's
crafty political director, and Club president Joe
Fontaine, a high school science teacher, decided
America needed a call to action. They started a petition drive to oust Watt. My classmates and I added a
few hundred signatures to the 1.1 million that were
collected.

Though I was eight years old at the time and
mostly concerned that Sharon might give me cooties,
I contributed to the pressure that removed one of the
greatest threats to the environment in history. James
Watt helped me discover that all of us, even a naive

kid in elementary school, can throw out greedy politicians. Sometimes I dream of meeting him and thanking him for his role in making me the president of the Sierra Club. I'd love to see the look on his face.

Even though I still had to ask for permission to ride my bike to the 7-Eleven, I could make a difference. Little did I know that the threats to our environment were not just in our national parks. They were close to home as well.

I loved to play T-ball. Unfortunately I played about as well as Michael Jordan plays baseball. The coach would stick me in right field far past where any kid could whack the ball. That suited me fine. The other kids couldn't laugh at me for dropping the ball. I spun around in circles and stared at the gophers as they popped their heads out of their tunnels to keep me company.

Every morning before I went to school, I grabbed my mitt and opened the *Los Angeles Times* to page 2 to check the daily smog report. In San Francisco, there's a weekly earthquake watch. In Providence, the papers print the Hurricane Watch in the spring. In L.A., where I wanted to run around outside in the famous California sun, we had a daily smog report.

A first-stage smog alert meant no T-ball, no playing outside, period. Some days the smog hung so heavily that the papers warned you not to "exert yourself too strenuously out-of-doors." I knew even then that there's something wrong when a child has to check a smog alert to see if he can play.

In the mid-seventies, first-stage smog alerts were common. The newspaper reported them once a month, and as much as once a week in the heat of summer. On "first stage" days I had to stay inside and watch *Gilligan's Island* to see if Gilligan would sink the coconut boat that was finally going to get them off the island.

When I reached high school, I sat in chemistry class bored out of my mind and stared out at Coldwater Canyon Boulevard from the windows. The road was only one hundred feet across. If I stared at the road long enough and unfocused my eyes I could barely make out the other side of the canyon. On "first stage" days, I couldn't see to the other side of the road. I sat in class struggling to understand chemistry, but aware that I couldn't see across the street. I didn't need to understand the science to know that something was wrong. I knew the air was poisonous.

Growing up in L.A., I suffered as much lung damage as if I were a smoker. I had no choice. Most kids didn't think it was strange that half of their friends toted asthma inhalers. Life in L.A. was sometimes like a Ray Bradbury novel, a surreal mock-up of what I expected.

Every school vacation my parents would pack my brother, Kevin, and me into the car for trips to the national parks. They wanted us to see that the world was more than L.A.'s smoggy canyons, nose jobs, and leased Mercedeses. We didn't go to see Siegfried and Roy in Vegas like many other L.A. families. We didn't go for the flash and glamour of L.A.'s celluloid spots. We went for solace. From the Grand Canyon to Zion to Yellowstone to Death Valley, from Acadia in Maine to the East Mojave in California, our vacations gave us respite from San Fernando Valley culture.

We were so struck by Yellowstone's awe-inspiring beauty on our first summertime visit, we visited Yellowstone a second time in the winter. We flew into Bozeman, Montana, taking a slippery bus ride to the edge of the park, and a trip by snow tracker to the one motel that remained open in the winter.

Never having driven a snowmobile before, I pulled all of my best Evel Knievel stunts—jumps from snow-drifts, fishtails on a frozen lake, anything I could do to piss off my mom. We didn't see another soul for hours.

Skidding up to a warming hut to regain feeling in our fingers, we saw thirty other snowmobiles parked outside the hut and wondered where they had been. It was as if all the other people just popped up through the snow at the huts and melted back into it when they left. After spending all my days in the sea of humanity that is L.A., I appreciated having a place away from the buzz of the city. Only a place like this could turn my attention outward to admire its strange and beautiful remoteness, and inward to contemplate my own solitude.

My parents created their own mechanisms to protect themselves from the vanity of L.A. As children of immigrants, both of my parents were raised to view anything new and modern as the true America, but they developed their own deep-seated connections to the old, the timeless, the wonders that never fade out of style. They chose to live on a ranch in the valley rather than in the pulsating heart of the city. My love for camping grew out of setting up makeshift tents in our yard, where my brother and I had a backyard of rambling, rough land for our imaginary games. Some of our neighbors kept their kids off their stiffly manicured lawns and neat rows of tomato plants. Our backyard was always for playing.

An old olive tree fills the center of the yard with gnarls of twisted growth, like the hands of a shriveled woman clutching her broomstick. Throughout my childhood I scrambled up its branches, digging my heels into the knots of its wood for support and hiding trinkets in its branches like Boo Radley. That tree taught me that you can never control nature.

Each spring, the olive tree fills with tiny white blossoms. They begin as buds, protected by a green leafy sheath that shows the white of the blossom only when the sunlight peeks through the branches at certain angles. Then the buds expose themselves with abandon. Their sweet scent fills the humming air as insects come to take their spots in the dance performance of life.

Then the olive blossoms drop. Everyone loves the

blossoms. My father brags to his jealous Yonkers friends, "It's L.A.'s version of snow, right in our sunny backyard."

Then an olive replaces each blossom, and the olive wars begin. While olives are great to eat when you have the time to pick and prepare them, if you leave them on the tree they eventually fall. As a kid it was my job to scramble around the tree picking off the olives. But with a tree this size, it is impossible to get them all. Falling olives make our patio into a war zone for three months. If we don't get bopped on the head, we trip on the rotting olives and find ourselves skating off the patio, shoes lubed by the juice of the fallen fruit. After family barbecues, relatives left our house cursing and holding their bruised elbows and knees. When I failed at getting all of the olives, my father tried pruning the tree, but the olives always came back.

My father—a doctor, a writer of reference books, a man of science—firmly believes that a solution exists for every problem. Like Bill Murray in *Caddyshack,* he was determined to beat his tiny foes. After careful consideration, he hired Jim, the local exterminator. Jim is totally dedicated to killing any living thing that annoys his customers. He's a professional.

My father asked Jim to "do the tree." When the band of flowery flashers threw open their coats, Jim rushed over to zap them. He even bragged on his first visit, "The flowers don't even have to open. My stuff here, it's so strong. It'll eat right through them

and destroy those little baby olives at their cores."
And so it did . . . at least for the first few years.

One year, my father called Jim just as the buds
had begun to open. Jim gladly came by to spray the
tree, but the little olives grew—and grew—and grew.
I knew we were in for trouble when the branches
hung low with thousands of little black balls. Each
olive pulled itself closer to the earth, waiting for its
moment of glory. Sure enough, the olives crashed to
the ground.

The next year my father took no chances. He challenged Jim with taunts like, "Olives are the returning
champs this year, Jim. Better be careful."

Jim told my dad that the problem, besides his
sense of humor, was that things had changed. No
longer could he spray when the white blossoms first
appeared. Unless my father wanted to waste time
and money, my father would have to wait until the
blossoms were opening before calling Jim. Not only
that, but Jim would have to make at least two house
calls, as different blossoms open each week. My
father felt a twinge of suspicion–two house calls
translates into twice as much money.

Knowing the joy Jim took in his ability to eliminate
pests, my father gave him the go-ahead, but not
without a few questions. "Why?" he asked. "What
happened to the great stuff you used to use?"

Jim reflected a moment. "Well," he replied, obviously upset, "they've taken all my good stuff away
from me. All they've left me with isn't worth a damn."

The government didn't take Jim's good stuff away from him to hurt him or to prolong my father's war with the olives. They took it away because it was too toxic. Most likely, they were protecting Jim's life from the toxic fumes of his "stuff."

Jim never wondered whether the new regulations were really for his benefit. He never thought that losing an effective spray could be more beneficial to him and his family than harmful to his business.

Given the choice, my father would choose the olive wars anytime over harming a single human being. I've grown to believe the government took away Jim's good stuff for one reason—the painful realization that sometimes, in our desire to subvert nature, we only succeed in destroying ourselves.

We like to be powerful. Our heroes always have superpowers or superabilities. Superman can change the course of nature, even rolling back time with the powers he possesses. Mythological gods hurled thunderbolts and razed mountains. We like to think that when nature gets in our way, we have an opportunity to show how powerful we are. We have a chance to become godlike. We miss the big picture.

I've learned a lesson from the olive wars as well. Instead of spending time and money trying to quell nature, I try to work with it. When it rains, I take my plants outside. When it fogs, it's a great day for photography. And when the olives fall, it's time for martinis.

How to Touch an Electric Fence Without Getting Shocked

In my junior year of high school, I left L.A. to spend a semester at the Mountain School in Vershire, Vermont. A small branch of Milton Academy, the Mountain School gathers forty-five students to learn in an integrated educational environment. The school is set in the living laboratory of a three-hundred-acre organic farm. Students take traditional courses like precalculus and English along with environmental science. While everyone else was taking precalculus, I was still trying to figure out what type of animal a hypotenuse was.

Everyone contributes to the upkeep of the school. Some tap trees to make maple syrup, some pick vegetables from the school's garden, and some bang nails to repair the roof of the chicken coop.

We were encouraged to skip a class, climb a tree, and write in our journals like Henry David Thoreau. Each student found an arena in which he or she led the rest of us. The student who rarely spoke in class gushed with ideas when it came time to design a footbridge. By watching one another excel we learned that each one of us had a role to play in the community.

When I first met the environmental science teacher, Kevin Mattingly, I thought I had met Einstein. He seemed to know all the answers, even if that meant the answer was "I don't know." I had never had a teacher who admitted to less than omnipotence. His classroom was typical, with one exception. One wall of the room was a picture window that framed a view of the auburn, maple-covered hills of Vermont. The window opened up onto the world. It was stunning, a constant distraction.

Kevin believed experiences were the most effective way to offer lessons. He once convinced a student to touch his nose to an electric fence to see if it would sting. It did.

Kevin had told the student, "It's an electric fence, but if you touch it with your nose, you won't get shocked."

The boy hesitated, his common sense telling him the teacher couldn't be right. He protested, "The fence is gonna shock me no matter which part of my body touches it."

Testing the mettle of the student, Kevin repeated, "You can touch an electric fence with your nose and you won't be shocked."

The student cautiously touched his nose to the fence and was rewarded with a shock.

It sounds deviant, but Kevin wanted the student to use his mind, not to blindly follow the words of someone who claimed to know the answers.

We need new answers, new ways to touch electric fences without getting shocked. When we are shocked by a river that floods its banks, we should learn to protect wetlands rather than build more ineffective dams that will shock us again.

On the first day of class, Kevin stood before us and explained that if any of us began to get sleepy, he had two rules.

"Number one," he sternly directed his words to all twelve of us. "If you get so tired that you have to go to sleep, don't pull the old head-in-hand trick." We knew we were in for it.

"I used to teach in a Catholic school where they had these old, hardwood desks packed tightly together." Kevin paused often while telling stories, to sift out each detail like a man sifting gold from the sands of a river.

"One day, one of my students, Brian, rested his head in his hand in a weak attempt at looking like he was paying attention while he was falling asleep. Brian didn't have the strongest arms"—another pause while he looked out the window at a grouse in the distance—"and as he drifted off to sleep, his arm buckled and slipped into the crack between the desks.

"Whack! He nailed his chin on the desk." Kevin slapped his hand down on his own desk. "I had to waste an entire day taking him to the emergency room rather than doing what I should, which is teach. So if you're going to go to sleep in this class"— the consequences hung like an H-bomb from the bottom of a plane—"just do it with your head down on the desk."

We were hooked. "Number two," he continued. "If and when you do put your head down, make sure you're facing that window." He pointed to a vista, hills rich with life, that some of us hadn't stopped exploring since we entered the room. "You can probably learn more from looking out that window than you possibly can from listening to me."

I had struggled in school. My first suspension came when I lit a match in preschool. During grade school, I was sent home for fighting a couple of times. I even failed in penmanship, leading the school to recommend that I see a handwriting therapist. In high school, I was a reasonably mediocre student. The Mountain School was the first educational experience that taught me to think. Instead of penalizing me for thinking in different ways, the faculty rewarded me. The Mountain School let the students lead; the teachers were coaches, not gods. Every kid should have the same opportunity.

I'm Just a High School Student

After my semester at the Mountain School, I returned to L.A. feeling like a house cat who had spent the day frolicking outside in the sun. I flew into Burbank airport through a smog cloud so thick that I couldn't see the city below. I closed my eyes and thought of Vermont's greens and blues.

That summer, I walked into a converted bank that served as the headquarters for the Big Green campaign in Los Angeles. Big Green was a sweeping state ballot initiative designed to attack foes of the environment on every front. The initiative pushed for cleaner air, more open spaces to enjoy, greater fuel economy, and the preservation of forests. It was the whole enchilada, a bold initiative.

The bank looked as if it had been stormed. Papers cluttered the floors and walls, hastily drawn maps were tacked to ceilings, and people moved frantically like traders on the floor of the New York Stock Exchange.

"Yeah, what do you want?" a gaunt woman barked at me.

"Well, I, uh, want to help." I figured that I'd lick envelopes until my tongue had a seizure.

"OK, what can you do?"

"I don't know," I replied, hoping that she would set me to a task. "I'm just a high school student. Who's the person organizing high school students?"

She looked at me with her feline, predatory grin and almost purred, "Well, you are."

"Me? I've never done anything like this before. I'm just a high school student." I said "high school" more slowly this time, figuring she was confused. I looked at her, waiting for some sign that she was joking, but she remained steady as she assessed me.

She collected some papers from the piles on her desk. "This is your chance. Now is the time to start. There's the phone. Read these papers. Get to work."

That gaunt women was Stephanie, a chain-smoking, stressed-out, sunken-eyed woman who looked as if she'd been awake for at least a week. She was intense and dedicated. I developed an incredible crush on her.

I learned everything I could about Big Green and started calling my friends to tell them about it. I didn't ask for much from them, just that they call the local office and offer help. We could write letters, walk precincts, make phone calls, do office work, whatever was useful.

Students all asked me the same questions I had asked Stephanie: Why me? What difference can I make? How will I know what to do? I told them what Stephanie had told me. "There's the phone. There's the coffee. Read these papers. Get to work." Before long, more than three hundred California high school students were canvassing door-to-door and staffing phone banks.

I had the time of my life. It felt more like a team than any sport I had ever played.

Grassroots organizers work harder than investment bankers and are paid like fryolator operators. We worked twenty-hour days and no one complained. Adrenaline, caffeine, and loud music kept the team alive. Everyone focused on winning. It was a rush.

After fighting the good fight, I figured that we had won hands down. Wrong.

We got annihilated, whipped, creamed, and had our toast buttered on Big Green, later nicknamed "Big Dream." I felt like Walter Mondale. The goal of the grassroots campaign was to change 2 to 3 percent of the overall vote, which we accomplished. We needed to change 8 to 10 percent of the vote to pass Big Green. The oil, mining, and timber industries ran television ads that reached far more people than our calls could. They outspent us ten to one—$30 million to our $3 million. Money talks. I listened and learned. You need an air war to go along with the ground troops.

I was discouraged. We had failed. I felt disappointed in myself for leading people down this dead-end path, and disappointed that I might have soured them on other battles in the future. I went to bed that night under the weight of all of these thoughts.

I woke up the next morning to the electronic wail of my phone. It was a fifteen-year-old girl named Tasha. "What's next?" she said anxiously. "I know we lost, but this was fun; it felt good."

"Tasha, I don't know. Let me call you back." I started to hang up the phone.

She didn't let me off the hook that easily. *"Adam!"* she screamed. I put the phone back up next to my ear. She continued, "Now listen. You're going to call me when you know what's next. I'm sorry we lost, but we've got to go on to something else."

The phone rang again. "Wake up, you bum!" This time it was Aaron May. "Don't tell me you're going to wallow. I hate wallowers. There will be no wallowing. None."

"Please, Aaron, let me wallow for a while," I begged.

"Nope. What's next?"

I had no idea. I started calling around for some help.

I called the larger sponsors of the Big Green campaign, groups like Greenpeace, the Sierra Club, and the Public Interest Research Groups (PIRGs). I asked each of them if they had programs in Los Angeles to channel this raw high school student energy. For different reasons, they didn't have a clue—Greenpeace wouldn't call me back and the PIRGs were trying to retain their college funding base. Finally, the Sierra Club's Los Angeles chapter invited me to its executive committee meeting to share my ideas.

My first impression of the Sierra Club was that everyone was old. I looked around for another young person and sat next to someone's six-year-old. He gave me half of his candy bar and we both sat at the back and watched. I tried to follow the argument they were having, but the words were unfamiliar and they all used acronyms like B.L.M., N.P.S., U.S.F.S, C.E.R.C.L.A., CONSCOMM, and HAZMAT. On the walls, black-and-white photos of bearded hikers ascending impossible peaks stared down at me. I went to the meeting with a foggy sense of what I wanted. They told me to come back in a month when I knew what I wanted to do.

I called students who had organized high school environmental groups and roped them in. A month after the initial meeting, I returned with a plan for the Sierra Club High School Environmental Leadership Training Program (or S.C.H.S.E.L.T.P. for short). I described the goal as creating a corps of student leaders who would bring the Sierra Club's mission to young people. Everyone listened intently. Then the wrestling began. The chair, Joyce Coleman, a dedicated volunteer with short red hair, asked, "Why should we go for this student thing instead of cleaning up the L.A. River?" The L.A. River activists grunted their agreement from the crowd.

"Look," I said, "If you approve this, we'll have fifty students to clean up the L.A. River next year. We'll start a chain reaction that will reinvigorate activism." I had no idea what I was talking about, but it sounded good. After brief deliberations they approved it, gave me $500, and we were off and running. I was amazed. They trusted me. They didn't even know me. I'll never forget the feeling. I was determined to live up to their faith in me.

I called Mark Fraioli, a more experienced student activist who had created and run the Lorax Environmental Club at Milton Academy.

"Hey, Mark, how would you like to help run a summer camp for the Sierra Club?" I asked, trying to sound casual.

Mark laughed. "What do you mean, 'run'? I didn't know the Sierra Club had a program."

"Er . . . well, I mean, how would you like to create and run a summer camp for the Sierra Club?"

At first, we debated the merits of joining an organization in which the average age of its members rose a year each year. The Club was mostly white and had a reputation for being a yuppie joint. The students we wanted to recruit weren't the golden retriever types. We questioned how the two groups would relate.

The Sierra Club used a different language than we did. What we called "activism," they called "conservation." What we called "hikes," they called "outings." You can't go to a high school and talk about "outings" without students breaking into homophobic giggling. Student activists abhor committees, preferring direct action; the Sierra Club had a Committee on Committees. Even the stickers the Sierra Club put in their direct-mail appeals looked more like baking soda coupons than what you'd put on your bumper.

Mark signed on. He was attracted to the challenge. Could we, a group of high school students, turn the nation's oldest environmental group into a youth powerhouse?

We started with the camp. It set out to provide fifty high school students with five days of excitement, support, and inspiration. It focused on organizing as an attitude of confidence and boldness. Mark planned to help the students minimize the same fears and insecurities that he had faced in founding the Lorax Club. Leadership is a lonely endeavor. Knowing that other people like you are trying to lead is the best motivator.

David Sabban, another staffer, headed our recruitment effort. He visited schools, trying to get biology teachers to convince their students to attend the camp. We didn't want the camp to become a suburban summer sports camp, so we recruited from some of the toughest schools in L.A.

Walking the halls of one school, he paused to peer into a classroom window to find a hanging skeleton or some other sign of a science class. Just as he stopped, security guards grabbed him. They whipped him around and pressed his face against the wall before handcuffing him and roughly leading him to a cruiser. David's crime was not having a hall pass. Drawing students from a school environment where you can be arrested for walking down a hall, we knew we had a challenge in front of us. The educational system was slowly beating independent action out of the students.

We had to figure out a way to show the students that they could lead. Their education so far had taught them how to follow. We looked for the class clown, the quiet nerd, the obnoxious bully—students who were already straining under the restrictions of their education.

Aaron May, a sharp-tongued friend I met while we were counselors in training at Camp Alonim in Simi Valley, California, planned a simulation event to prove to them that they could successfully implement a campaign. Each camper had her own role. Some created budgets, some organized rallies, some lobbied, and others fed the media with creative press events. As counselors, we acted the roles of industry interests, politicians, reporters, and funders, to give the students the feeling that they were actually involved in campaigns.

"We wanted to challenge them and make them think that the task was too much," Aaron remembers. "That way, at the end, they would realize that they could do much more than they thought they could." The students learned practical skills and how to apply them without being limited by their insecurities.

We were young and inexperienced; that was our strength. The students trusted us. We called ourselves staff members even though no one was paid. We were students leading students and it showed. To look older, we all carried clipboards. I grew a beard, thinking it would make me look respectable. (It just made me look fuzzy.) The oldest staff people had just graduated from high school. There were students at the camp who were in college.

At the end of the camp that first year, the campers gathered around a bonfire. It was a tense time for those of us who had invested so much of ourselves in the success of something unknown. Everything seemed to have gone well, but now the jury was in to deliver its decisions after a week's deliberation. One student, Cindy, talked about her plans to start a new county student environmental group. Another student outlined a lead poisoning prevention campaign. Each student spoke about plans for the next year at school. They were ready.

"So, you leaders, what do you want to do now?" I asked.

They all responded in their own ways, each saying the same thing. "We want to go further with this. We want to take this and bring it home and share it with the other students we know."

"Well, what do you want to call it?"

After ditching some early attempts like "the Student Majority," "the Sierra Cubs," and "the Yuppie Solution," we agreed on "the Sierra Student Coalition" . . . the SSC.

A week after the bonfire, the counselors met at the Los Angeles Chapter of the Sierra Club to start networking. After we graduated from high school, we took our energies to college. Eventually, Mark, Aaron, and I ended up at Brown University together where we formed the hub of the Sierra Student Coalition. We started working with students across the country who were looking for some connection to a larger movement.

When Vice President Dan Quayle attacked *Murphy Brown* for its lack of family values, we organized screenings on hundreds of college campuses of the *Murphy Brown* episode that responded to Quayle's criticisms. We used the events to register students to vote. At Brown, seven hundred students packed Solomon Auditorium for the show. We started with a "Quayle-roast" where we read off Quayle's most famous goofs from a podium filled with sacks of potatoes. Quayle had just misspelled the word *potato* at a grade-school spelling bee.

When the California Desert Protection Act (CDPA)

came to a vote in Congress, the fledgling SSC began a new technique that we called the "Dorm-Storm." The Sierra Club's legislative office E-mailed the campus leaders and told them the target. The students called together the other dorm storm troopers and moved quickly down the halls of the dorms, telling each resident to make a call or send an E-mail to a member of Congress who was "gettable." Working with Club activists across the country, we nailed down enough senators to help pass the CDPA and create three new national parks, including Death Valley National Park, the largest in the lower forty-eight states.

Meanwhile, the national office of the Sierra Club had been hearing rumors of this "student thing" and began to investigate. We were summoned to report to the board of directors immediately. We had blatantly broken Sierra Club policy and procedure. We had skirted the rules. We had started a national program without permission. We had registered 100,000 eighteen-to-twenty-four-year-olds to vote without anyone's instructions. How irresponsible.

We went to face our fate as a united force. The board of directors was deeply divided. The oldest members of the board quickly embraced the idea of encouraging students to push forward the environmental agenda using our own language and fresh, new ideas. Feeling threatened by a structure that they did not help create, some of the aging baby-boomers plotted against us. Someone had climbed into the treehouse without knowing the secret handshake.

I remember the debate being funny and sad. "We simply can't approve an organization called the Sierra Student *Coalition*," one of the Club's fifteen directors declared. "*Coalition* is a word that implies the legal binding of two organizations. We'll open up all sorts of liability. Those kids could go out and blow up bridges or spray-paint fur coats and the Sierra Club will be left holding the bag. . . . They're kids, loose cannons." I shook my head. People always argue procedure and liability when they don't have the substance.

Another director expressed her opinion of the SSC, saying, "Adam, I don't know how we can afford to have a student program. We can't even afford a secretary to take dictation from board members." We needed funding to connect students to their local chapters and to set up a central office for the student program. Besides, it didn't take a soothsayer to see that the student activists of today are the major donors of tomorrow.

"Well, if that's your priority then I guess I'm with the wrong organization," I responded, amazed.

Here we were, offering energy and excitement, and some board members wanted to talk about liability and dictation. I felt ready to quit.

In the heat of the debate, Ed Wayburn, who had been silent throughout, raised his solid but soft, eighty-seven-year-old voice. Wayburn, the honorary president of the board, cleared his throat and said, "I'd like to make a motion . . ."

That's as far as he got before the elected president

of the board silenced him, saying, "You can't make a motion because the motion on the floor is to dissolve the Sierra Student Coalition."

Undaunted, the honorary president returned, "I'd like to take a moment of personal privilege." The chair nodded. There are some advantages to age. He continued, "My personal privilege is that I'd like to make a motion to congratulate and thank the Sierra Student Coalition for their work in bringing new energy to the Sierra Club." He looked around the room at the board to rally the supportive members. "I'm sure that the board will support this program," he ended. The opposition crumbled. It was the oldest person in the room who saved the youngest.

The SSC made a clear statement to the larger Club that young people were ready to take responsibility for the world they inherited. The Sierra Club offered the SSC a strategic vision and the wisdom of experience. The student movement had never lacked energy—the Sierra Club gave it focus.

In the 1980s, the Sierra Club had moved toward an insider's approach to activism. Most chapters had five or six well-informed volunteers who worked together, spent weekends together, discussed the issues, and acted as a closed unit. Their style involved educating the few to deliver the message to congressional targets. The national Sierra Club's response to President Reagan's ongoing assault on the environment was to hire a conservative Republican executive director to refine the inside game.

In the early 1990s, the Sierra Club began its rebirth. Carl Pope, formerly the crafty political director, had been promoted to executive director. He immediately began focussing the Club's efforts. The board of directors reorganized the seventy-three issue committees and streamlined their management structure. A member's campaign to halt commercial logging in the national forests pushed the board to take stronger positions. The chapters adopted strong environmental justice programs, and the SSC empowered activism in anyone we could reach.

Following the Sierra Club's mission requirement of using all legal means necessary, we did anything we had to do to get young people involved. The SSC arrived just in time for the greatest congressional threat to our environment, in November of 1994.

After the people voted and their votes were tallied, America sent the 104th Congress to Washington with Newt Gingrich as the Speaker of the House. The new congressional majority threatened to drill for oil in the Arctic National Wildlife Refuge, destroy Utah's wilderness, and even emasculate the Safe Drinking Water Act. The students in the SSC grew tired of letter writing and petitions. They began brainstorming for ideas to express what Congress planned for our earth. Piñata Day was born.

Michelle Sheaffer heard about Piñata Day through an SSC mailing. This was one action that wasn't for procrastinators. The day before Piñata Day she pulled an all-nighter to prepare. She began making a giant Earth-shaped piñata by slopping strips of gooey, mucous, glue-soaked newspaper onto a balloon. After three layers, she left her creation to dry.

Returning hours later, Michelle found the balloon still soaking wet. In a panic, she blew it dry until her hair dryer's motor burned out, but the plastered balloon was still dripping wet. Determined to dry out her piñata, she enlisted the help of an oscillating fan. Michelle went to sleep to the sound of the fan, its rhythm calming the excitement she felt for the morning, now just three hours away.

The next morning, Michelle walked along the pathway toward the Mall in Washington, D.C., feeling alone. There weren't many other kids around. When she reached the booth and set up her piñata, the Sierra Club members already there were nervous about her strange concoction. They told her to stand far behind the table where no one could see her. She started to move her globe out of sight, and then remembered what she was there to do. She hung her globe up right in front of the table. She smiled and picked up a black bandanna marked with the name NEWT in glitter. Handing a Louisville Slugger to her three-year-old nephew, she put the NEWT bandanna over his eyes, picked him up, and told him to start swinging.

NEWT sparkled in the sunlight as the boy started bashing the piñata and Michelle began shouting, "This is what the 104th Congress is doing to our Earth! Tell Newt to stop bashing our Earth and my future! Save our environmental laws so my nephew and I can drink clean water and breathe clean air!"

People stopped and huddled to see what the fuss was about. The Sierra Club volunteers were awed by the crowd. Parents let their own kids take a swing at the Earth-shaped piñata. After thirty people had their shot, the piñata broke and the kids scrambled after the toy skeletons and soldiers that had rattled inside.

Michelle carried her broken Earth back through the crowd and watched people walking around asking for signatures on petitions. She had garnered fifty signatures from the parents of the bat-swinging kids and had fun doing it—much more than just asking for signatures on a petition. Everyone would remember her image of Newt bashing the Earth.

Today, the SSC distinguishes itself from other student organizations by its connection to the Sierra Club. Old fogies like me move on when we graduate and let new students run things. The SSC benefits from the focus of the Sierra Club's campaigns and the Club retains the creativity and fun of youth. The environment wins.

Signing My Life Away

I was frustrated by the initial response of the board of directors to the idea of the SSC, so I ran for the board. Any Club member in good standing can run. The members elected me. During my first year on the board, I was elected as the fifth officer of the board. What is the fifth officer? To this day I have no idea. In my best estimation it was an organizational place holder to fill out the executive committee of the board—no power whatsoever. After I was elected, David Brower, the legendary conservationist, chuckled at me, "Just don't forget, Adam, it's the fifth wheel that drives the car."

I used that year to learn the ropes of the organization. Two years later, the board huddled together and decided to give me a shot as president, making me the first Sierra Club president for whom Pong was a life-changing event.

My agenda is simple.

1. Shift the focus to the grass roots. They're the heroes. Teach neighbors the tools they need to hold large institutions accountable.

2. Reach people the way they learn. Never assume that everyone understands your point. Use the media to spread the message.

3. Fight for your values; they're not negotiable. Don't compromise on the basics—clean air, clean water, safe food, biological diversity, and the protection of our last ancient forests and wild places.

4. Be young. Have fun.

In the last two years, we've shifted 80 percent of the money we once spent on direct lobbying to grassroots organizing and outreach. The strategy has worked. We've won major battles to protect clean air. We've saved the American River from being dammed. We've preserved 1.7 million acres of Utah's wilderness. We've secured funding to restore Yosemite. We've protected the Okefenokee swamp. We saved the Sterling Forest from becoming condominiums. Seventy percent of Sierra Club–endorsed candidates were elected in 1996. I traveled 150,000 miles in my first year as president to work with local Sierra Club groups on their agendas. We've hired fifty new grassroots organizers across the country to get America riled up and ready. With the release of our first CD, the Sierra Club will be using music to spread the message.

As president I also have to take care of a lot of details, which are often as telling as the big picture. Many documents require my signature: letters, reports, responses to angry polluters, checks. The list unravels like a roll of toilet paper. When I took the job, I figured I'd be tagging papers about 90 percent of the time, but a signature machine does the work for me. It's ironic that signing my name was the way I first got involved in the Sierra Club.

In order to make a signature machine work you've got to give it a signature. I tried. And I tried and tried. But the signature machine wouldn't accept my scribble. The machine couldn't imagine that the squiggly line I submitted was an actual signature.

They asked me to change my signature. I asked them to change the machine.

It was an inauspicious beginning. I thought I would fit nicely enough into the structure to change it quietly. I knew it was going to be a hell of a ride.

SECTION ◇②

*Good Kids Draw
Outside the Lines*

Construction Paper

I always get asked the same question. It's really more of a statement than a question, but people always phrase it the same way. "Aren't you too young for your job?"

My reply is the same every time. "No, I'm too old. A fifteen-year-old would get more done."

Kids don't worry about drunk drivers, sexually transmitted diseases, making a commitment, paying rent, missing deadlines, and eating the same dinner of Kraft Cheese and Macaroni every night. They haven't memorized the boundaries of what's acceptable. They don't know what "won't work." As we get older we begin to accept the unacceptable. We accept that bad things happen. We rationalize. We follow too much advice. We seek therapy.

We need to rekindle the youth within us.

The older we get, the less encouragement we find to experiment and wander beyond the boundaries of a constructed world. The paper on which we write is a perfect example. In preschool, we're handed blank sheets of construction paper that we fill in any way

that pops to mind. We grab our crayons and begin.
Drawings of crazed fuzzies spill over the edge of the
paper with no respect for boundaries. We glue buttons and pieces of yarn on the paper to give it tex-
ture, an added dimension. We fill it up, or out, or all
over, reigning freely over the blank space. No sug-
gestion is made as to whether we fill it with pictures
or words. All paths are free and clear.

As we move up a grade or two, the blank paper
develops a line at the bottom. The line looks more
like a racetrack, with bold lines on the outside and a
thin dashed line in the center. The "unstructured"
work done above cannot stand on its own, free of all
rules like the contruction paper of preschool. Any
drawings must fit into the blank part of the paper,
and the racetrack line must be used to explain art
with words like "dog" and "house." A simple line
establishes a simple rule for the use of the paper.

Move ahead another year, and the whole paper is
filled with big, wide lines. Rules are established.
Some leeway exists in how the letters are formed,
their shape and size, but the lines direct us only to
use letters. No more drooling iguanas or three-
footed cranes. The rules are clear. Stay within the
lines. Communicate in one specific way. Drawing
becomes something that you do in art class and summer camp. Only the talented artists continue to use
blank paper.

As we get a little older, the lines move closer
together. Teachers accept book reports only on wide-

ruled white paper, with your name, your homeroom, and the date in the top right-hand corner. Then, finally, you move to college-ruled paper. You have to follow "college rules." No words may be scribbled out, no large spaces between the words. Your two-page essay answering the specified question is due on Monday at 8 A.M. Lines narrow and structure creeps in.

Eventually, all of your work must be typed on a computer, single-spaced, with standard margins, in reports of no more than twenty pages and no fewer than eighteen.

I miss the days when I had the opportunity to express myself with drawings as well as words. I wish I had become confident enough in my drawing skills to be comfortable drawing what I feel. I find that emotions are better communicated by pictures and photographs than by writing. Perhaps that explains the success of graphics-heavy publications like *People* magazine. We all start school with a chance at graphic ability. Eventually, except for a few talented artists, we are weaned from drawing. Art is considered frivolous for serious communication. Ask anyone in advertising whether graphics are frivolous in their communication objectives.

Groups interested in social change have relied on the power of their ideas rather than the form in which they are communicated. That's why many of those "good ideas" sit on a shelf collecting dust. We can learn a great deal from kids. It's time to reclaim the skills of our youth and to pass on the challenging message of hope. We need to draw. We need to get down and dance together. We need to emulate the kids who avidly raise their hands to volunteer for any assignment.

Taking Newt by the Horns

I used to love raising my hand in class. I liked raising my hand so much that I volunteered for audience participation every chance I had. I lived to be chosen for Zamfir's "Saw the kid in two" trick, or Zippo the Clown's "Let's douse the kid in Jell-O" gag. My mother never appreciated this tendency. As I got older, I stopped volunteering for things that got me laughed at.

One of my favorite "hand raisers" is Rick Takeda. Rick is a surfer who hustles along the beach to catch the killer morning waves. As the stereotypical California surf bum, he makes an unlikely candidate for the role of activist. He amazed even himself by speaking out alone.

At Colgate College, Rick took time off to work on a 142-foot square-rigged sailboat and research station. In a remote part of the Caribbean Basin, a place far from oil rigs and heavy industry, they discovered tar in lumps the size of baseballs. The tar shouldn't have been within five hundred miles of where they were, but the garbage from the U.S. had already reached the basin.

"I had never been educated on pollution or the environment," Rick remembers. "I was from a conservative family, born in a conservative area. I was blown away that forty miles from land, in this pristine area, you could find tar the size of baseballs. That spurred me on to start looking at the environment in my own backyard."

By the summer of 1995, Rick had spent six months collecting signatures for an environmental bill of rights as a response to the Republican assault on the environment. Rick learned that Newt Gingrich was holding a book signing in Boston. He gathered forces, enlisting a local labor group and a women's group to protest Newt and his new book, *To Renew America*. Together they handed out three thousand bookmarks, one side printed with Gingrich's voting record and the other with the following quote from his book, "We have an obligation to minimize damage to the natural world."

Rick went to the signing with a copy of Gingrich's book tucked under his arm. He waited in line behind the men in suits. He looked around to see the conservatives of Boston, with platinum hair to match their platinum credit cards, queuing up so they could praise the Speaker of the House for his efforts to reduce regulation and protect the interests of the American people.

He nervously approached a table covered in books fearing that Gingrich could point a finger and disintegrate him on the spot. "I expected a forked tongue

and horns on his head," Rick remembers, "but he
just looked pudgy. I swallowed any self-doubt
because I had a mission."

As Rick handed him the book, Gingrich snatched it
with a doughy hand and, without even looking up,
began to sign it. He got as far as "Dear Rick" when
Rick cleared his throat and began, "Speaker
Gingrich, I just want you to know that we're going to
drop one million signatures on your doorstep to tell
you to stop messing with the environment."

Gingrich finally looked up and, with vacant eyes,
replied, "One million signatures, huh?" as if the paper
boy had just threatened him with tossing the morning
newspaper on his doorstep. Little did Gingrich know
that the environmental bill of rights would be a crucial step in coordinating the opposition to his war on
the environment.

Rick had to step up to the plate to find out the
pitcher had no heat. He was nervous about his
encounter, but that didn't stop him. He had never met
the Speaker of the House. He had never coordinated
a million-signature petition drive. He just swallowed
hard and gave it his best shot.

Sometimes you'll make a fool of yourself when you
try. Eventually you'll dust yourself off and get over it.
More often than not you'll get the job done—but only
if you have confidence in your own abilities. Stepping
up and making your voice heard is just like drawing
outside of the lines. It's fun.

Generation X . . . Don't Do Me Like That

Despite the efforts of young people like Rick Takeda, my generation has grown up under the derisive label of *Generation X*. Author Douglas Coupland cemented the label of Generation X to young people with his wandering cultural novel of that title. He tagged youth with a materialism equal to the BMW-driving brats of *Beverly Hills, 90210*.

Coupland described Generation X as being under-employed, stuck in McJobs, and suffering from malaise, its motivation stifled by cynicism. All that its members seem to be interested in, Coupland implies, is money. Throughout society, in movies from *Slacker* to *Reality Bites,* young people are told that they are apathetic. If you continue to tell someone that she's apathetic, she will eventually fall prey to your words, becoming frustrated and cynical.

Generation X came of age listening to Nine Inch Nails scream about angst and watching reruns of *Family Ties,* reflecting the Reagan-era economic phi-losophy of "Take what you can and then some." Alex P. Keaton symbolized pathetic, heartless success. Although he excelled in all areas academic and finan-cial, it took him years to develop enough emotion to hold a girlfriend or learn to relate to his family.

The Day After, a TV miniseries, filled the night-mares of the eighties with cold-war visions of nuclear holocaust. AIDS took friends in seemingly hopeless battles for life. Date rape fanned the fires of fear while I was at Brown, leaving girls afraid to go to parties and boys afraid that their names would end up on a list of rapists in a bathroom stall. Meanwhile, at Harvard Business School graduations, students waved one-hundred-dollar bills to celebrate their entrance into the world of business. The media reported failed S&Ls, molesting priests, and political scandal.

Generation X was shaped by the degeneration of the world it inherited. Our birthright is environmental devastation and a crumbling educational system—thank you very much, Mom and Dad. The same boomers who are responsible for many of the problems that Generation X faces lambaste us for not being as socially active as the hippies of the 1960s. The myth of 1960s activism has been overblown. More young people are involved in community service today than in any previous time in American history.

The members of Generation X are bitter about having to pay the skyrocketing costs of an education. The American Council on Education reports that in 1969 43 percent of all college students worked outside jobs while enrolled in school. By 1979 that number had jumped to 51 percent, and by 1990 it had jumped again to 63 percent.

I worked as a handyman at Brown. I've worked as a lounge singer in Japan, where I sang for drunken guests at Japanese weddings. I once enrolled in a medical experiment that admitted me to a hospital for a week and fed me No-Doz. Then they intravenously pumped me full of hypertension medication and took blood samples from me every five minutes to see how the medication worked. I earned $1,100 for the ordeal, which paid my credit card debt and infuriated my parents.

Watching heroes fall prey to temptation and seeing radical, morally motivated hippies turn into Range Rover–driving yuppies with a "me first" attitude could make McGruff the crime dog as uninterested in society as a serial ax murderer.

This was validated by a March 13, 1997, *New York Times* article that "confirmed" that Americans no longer care about politics. To prove the point, a reporter traveled to Michigan, a pivotal electoral state. He set out to determine why Michigan residents, and Americans in general, reacted to the newest Washington uproar over campaign fund-raising with a shrug. In the article, Bob Teeter, a Republican campaign consultant, commented, "Every person has only so much attention to give and politics and government takes up only a fraction of what it did twenty-five years ago. Look at the declining television coverage. Look at the declining voting rate. Economics and economic news is what moves the country now, not politics."

The new president of the University of Michigan, Lee Bollinger, was quoted, saying that on college campuses "social idealism is dead or dormant, and a vast majority see the United States as king of the hill, at least for a while." Bollinger continued, "In my mind, the absence of a taste for scandal is directly linked to the lack of a social agenda. If you have no idealistic vision of your country, there is no reason to be disturbed by leaders who fall short of high ethical standards."

Bollinger, the president of one of our country's largest public universities, a man who supposedly directs the energies of our youth to better themselves and their world, suffers from a belief in the myth of Generation X. It seems strange to entrust young people to someone who has so little faith in them.

That's exactly how University of Michigan student Ami Grace felt. She let Bollinger know her feelings in no uncertain terms. "As student activists, we feel that [your] statement was inappropriate and a misrepresentation of the University community," she wrote.

"The University of Michigan has over six hundred student groups on campus," she continued. "A large percentage of these groups are committed to serving the community and working for greater social justice. Their student leaders and members expend a great deal of time and energy toward these ends."

She showed the accomplishments of particular student groups, focusing on Voice Your Vote, an active student group on campus. "This non-partisan group was formed for the purpose of promoting and increasing voter registration. During the 1996 elections, they regularly set up voter registration booths and campaign information sites throughout campus. They also sponsored a political community service day that placed student volunteers in over fifteen different political organizations. Not only did Voice Your Vote register over 6,500 students in three months, but they also succeeded in producing one of the largest voter turnouts on a college campus in the country." Not bad for an election year that saw youth participation fall below 35 percent.

Ami was not content with simply proving Bollinger wrong, though. She was prepared to tackle the problems facing student activism in the school community: "We do not deny that student apathy exists. It is a problem all groups must face when trying to mobilize others toward some action." She continued, "These groups, although rather successful in advocating social awareness among some students, are often hindered from expanding their efforts to reach more students by a lack of funds or other resources. We believe that this is partially due to the less than adequate support that has been received from the University administration."

Ami concluded her letter by suggesting that Bollinger meet with her to discuss the problem. "By working together, we can better recognize existing student efforts by student groups on campus while providing a pathway for increased activism."

Ami then plastered the campus with copies of "Bollinger's B.S." Bollinger called her for a meeting the next day.

Unlike previous generations of Americans, Generation X suffers from limited expectations. We've lived all of our lives with scandals in the media, national debt, and unsafe air. If we hope to have a positive future, we need to break young people out of this mind-set of mediocrity. Young people need to demand a future that is better than the past. The good news is that Generation X is heading in that direction.

An Ocean and a Starfish

If presidents of universities don't support and encourage young people, young people need to take care of their own. There is no time to wait for adults to get a clue about what young people need to get active. I remember the preachy high school assemblies that were supposed to make me socially conscious. The only thing more boring than my high school assemblies would be John Tesh explaining where he finds his musical inspiration. Ocean Robbins has turned school assemblies around with Youth for Environmental Sanity (YES!).

"Our first performance was at Galileo High School in the gymnasium," he remembers. Their sound system for the performance consisted of a single megaphone. "The acoustics were so bad," says Ocean, "that if everyone was totally quiet they just might have been able to hear us." Three people clapped when they finished.

Looking back, Ocean reflects, "It was a miracle we continued at all. They couldn't hear us, literally or figuratively. It was a rude awakening."

But the group refused to give up. "After that we stopped being teachers and increased the entertainment value, increased the heart," Ocean said. "Instead of a message focused on data, we focused on what it's like to be a young person in this crazy world. We talked about our lives, how environmental problems had affected us. We added more theater."

When they performed at a high school in Centerville, Ohio, they had the routine down.

"Stereotypes are like scum on a window. The scum clouds the view," the presentation began. Then, back and forth, they recited an alphabet of stereotypes—A is for airhead, B is for bimbo, C is for communist, D is for druggie, E is for environmentalist, and so on. They loosened up the audience.

Next, a young woman named Ivy skipped onstage wearing flowers and a wildly printed skirt. After a few twirls, she gave the peace sign to the students. "We just have to open our hearts and love each other," she implored. "If Jerry had one message for us it's that we're all one tribe. Peace, dudes, sisters." The audience cheered and laughed.

Rachel strode onto stage next. Her hair cascaded from a pink scrunchy at the side of her head. She was the Valley girl, in miniskirt and Ray-Bans. She started with a whine, "Oh my God, like, it's sooo cool to save the Earth and be an environmentalist." Laughter again spilled from the audience.

Ocean, dressed in green polyester floods, fidgeted with a pair of broken specs and pulled pens from his pocket protector. "According to scientists," he paused, warming up his best Urkel voice, "ozone in the upper stratosphere is interacting with oxides in a way that is slightly different than was originally perceived in the mid-1980s."

Ocean and the members of YES! never took themselves too seriously. "We had to make fools of ourselves," he explains. "We had to play out everything that people thought we'd be. We became the solvent we needed to clean the scum on the window. The truth is that we're not any one of those things. We're young people, we care about our world, and we care enough to do something to help."

YES! has reached more than 550,000 high school students with its message.

Another YES! presentation unfolds in another high school auditorium. This time Ocean Robbins is an old man, with glasses dangling from his nose and a cane in his hand. He stands and watches as a young girl walks toward him along an imaginary beach. The night before, a huge storm crashed along the beach, leaving hundreds of starfish stranded beyond the reaches of the tide. As the two humans enjoy the sunshine and calm of a poststorm day, the starfish start to fry in the sun.

Walking along, the little girl picks up starfish one by one and tosses them back into the ocean, saving their lives. She approaches the grumpy old man, who is tired from a life of battling foes he can't see. He turns his gaze up from the sand to watch her and, when she comes within earshot, cackles at her, "Little girl, what do you think you're doing?"

A bit surprised, the girl looks him in the eye and replies, "I'm throwing starfish into the water to save them." Her big blue eyes disarm the cantankerous man for a second. She picks up another starfish and tosses it to safety.

He shakes his head and says, "But, little girl, there are millions of starfish up and down this beach. There's no way you can possibly save them all. What you're doing doesn't make one bit of difference." He coughs, clearing his throat, and delivers his final edict: "It doesn't matter!"

The little girl, unafraid of criticism, reaches down, picks up another starfish, and tosses it into the imaginary ocean of the auditorium. She smiles at the old man and says quietly, "It mattered to that one."

By throwing one starfish in the water you can make a world of difference. Ocean, Rick, and Ami are three examples of the thousands of young people throughout the world who take an active role in creating their future. They are the antidote to hopelessness.

SECTION ◈③

The Many Faces of
Environmentalists

And Now, Introducing the Environmentalist LX240

*T*he modern communicator takes full responsibility for his message. If the target doesn't understand the message, it's the fault of the communicator. I'm sick of environmentalists who preach as if they're trying to convince themselves. They bask in their own egos, refusing to understand that different people have different perspectives.

In reality, many species of environmentalists exist, all with different agendas and all with different driving forces. Most people have a few different species within them.

"Druid" environmentalists, driven by a basic belief that nature is restorative and inherently important for its spiritual qualities, have saved Yosemite, the great wilderness in Alaska, and other national parks across our country. For them, nature must prevail so that other generations may enjoy its splendor and be healed by its powers. They look to texts like John Muir's *My First Summer in the Sierra* and

Edward Abbey's *Desert Solitaire* for inspiration.

"Polar-Fleecers" base their concern on preserving our natural heritage as their wild playground. They focus on keeping rivers clean and the fish in them abundant. They look to limiting the use of nature so that it can offer recreational opportunities for years to come. They are rafting, swimming, hiking, and climbing through the wilderness in an attempt to entertain themselves, enjoying the ultimate amusement park. You'll see them sitting uncomfortably in a city, reading a Jon Krakauer book.

"Apocalyptic" environmentalists alert us to the ways in which we destroy our planet and ourselves. They map holes in our ozone, chart the effects of pesticides on children, explore the relationship between forests and oxygen levels, and look for alternatives to habitually destructive practices. They urge us to take action. Their concern centers on the ways in which we can survive as a species and avoid the mass suicide that cynics believe is inevitable. They look to Bill McKibben and Rachel Carson books for inspiration.

"Eco-opportunists" range from lobbyists to representatives to judges. They work within the political system to effect change, and they relate to the language of laws and riders. Although some genuinely bring the concerns of their constituents to the political arena and work to address the problems of communities, many simply come with the handi-election-environmentalism kit. These cucumber congressmen, green on the outside and seedy on the inside, push the message of hope while acting on different convictions. Some eco-opportunists use the political system to protect the environment; others use the environment to conquer the political system. They read Dixie Lee Ray and Julian Simon to confirm their cynicism.

"Eco-entrepreneurs" see the financial reality of dwindling resources. A paper company can't produce paper without trees if all of its machines are set to process wood fibers. A chemical company like Dow Chemical can't turn a profit if consumers boycott its many products in response to its disregard for our environment. The issue is numbers, and it's simple. It's reason enough to become conscious of our environment and its needs. Paul Hawken's *Ecology of Commerce* is their bible.

I don't care why someone cares about the environment, only that they do. Although poll after poll shows that more than 75 percent of Americans see themselves as environmentalists, the environment rarely ranks among the top five national problems according to public opinion. If you ask people about local problems facing their communities, the environment almost always ranks in the top three.

The leadership of the environmental movement tends to be most heavily populated by Polar-Fleecers and Druids. At the Sierra Club, we run more than twenty thousand excursions each year into natural areas around the globe. Hardly a day passes when one of our staff members isn't skipping off to Mount

Tamplais, a grassy peak with spectacular views of the Pacific and San Francisco. Bruce Hamilton takes his daughter Katie to river-kayaking lessons at night. They practice water rolls, spinning the kayak like a hot dog on a spit and submerging themselves in cool waters.

Druids tend to be more radical than Polar-Fleecers. They believe nature should be protected simply for nature's sake. A Druid is defined as someone who sacrifices people for trees. They are the warriors who block commercial use of thousands of acres of wilderness so that a dwindling species may survive. For them, nature need not be saved for humankind's use. They believe that nature deserves to live for its own sake. It is the language of Druids that dominates the environmental movement and often alienates potential allies with its focus on the importance of nature above people.

Whatever the motivations, as long as each group of environmentalists works toward the same goal of conserving, protecting, and restoring our natural heritage, they are all welcome. Whatever your motivation, allow people to have their own. We can attain higher goals with a milder message.

Clara

For Clara, and millions of kids like her, the environment is not about buzzwords or legislation. It's about running a mile.

She throws on her lightweight nylon shorts and tank top on a sunny February San Francisco afternoon. During her day at school, all she could think of was the moment when the bell would ring and she would be free to run. Her body bears the mark of well-nourished youth, with muscles beginning to take solid form and skin that shines dark like the depths of a cool lake in pale light. Looking at her provides a glimpse at what could only be called perfect health.

Like the environment she lives in, Clara appears to be running strong. But, like the environment, her appearance of strength is misleading. Clara has chronic asthma. If she doesn't take care of herself, she will lose her most basic ability to function.

Little particles, bits of solid and liquid that are so small that thousands of them could fit on the period at the end of this sentence, pick at Clara's respiratory system. They're known as particulate matter, and they are formed primarily by fuel combustion—diesel in trucks, wood in stoves, gasoline in lawn mowers, propane in the grill. The particles find their strength in their tiny size. Because they are so small, they succeed in penetrating deeply into the most sensitive areas of the lungs and respiratory tract.

Particulate matter aggravates Clara's condition. She is one of America's children. She is twelve. She hopes to go to college one day. But right now, Clara would just like to run a mile.

Clara competes in the 200-meter hurdles, which is a step beyond the 50-yard dash, her first competitive event. By using her medication and being careful not to run on smoggy days, she has expanded her range. When smog and soot clog the air, or when she's caught downwind of a smoker, or when she walks through the black exhaust cloud of a bus, her goal of one mile is pushed back as if it were an endless marathon.

Clara isn't bitter, although she has every right to be. With eyes spread wide, exuding every confidence that she would succeed, Clara explained to me, "I am taking care of myself. I am practicing my new breathing exercises. I keep my asthma inhaler nearby. I am learning to deal with it, but I can't do it all the way alone. The air is too bad."

There's something wrong when a child can't reach even Clara's simple goal. Clara's fight is one she wages against an unknown foe, and one in which the odds lie heavily weighted against her. Hers is not a competition she willingly joined. Circumstances beyond her control thrust her headfirst into a game of chess against Bobby Fischer. There's something wrong when a child can't run a mile.

When my visit with Clara ended, she reminded me, "I will run a mile, and maybe more."

The Holy Ghost Slap

Children bear the brunt of our environmental irresponsibility. Their dreams are stunted. Their fragile health is sacrificed. Their chance at a safe start is compromised. In West Dallas, Texas, the kids learn to grow up fast.

The Reverend R. T. Conley, pastor of New Waverly Baptist Church in West Dallas, speaks with the pacing of considered thought. "They call me the Rev," he told me with his heavy drawl.

The Rev is Texas tough. He began his active career in environmentalism in 1969, eleven years before becoming a pastor, when the lead from a smelting plant in his town caught his attention.

"I lived here all my life," he explains. "I didn't know that the lead smelter was a health hazard; I just knew it did something to my house and to my car."

The Reverend Conley started asking questions about the lead smelting plant when he noticed that the color of his car was fading quickly and the paint on his house was peeling off inexplicably. "I went to the city complaining," he remembers, "and they said it was no problem. They told me to go away."

Rather than accept this brush-off from the city, the Reverend Conley went to his community to get them involved. He started with the pastor of his church, but he was shrugged off. People did not understand the health effects of lead at the time. "So I started organizing the community," he said. "I did a lot of legwork, knocking on the doors, asking them if [the lead] had done any damage to their cars. We checked each car. You see, for broke-down cars it would take all the color off the car. The side of the car facing the plant would be rusted out." Conley later found that the smelter's owners knew the effects of lead on the cars and possibly on people's health, but failed to tell the community.

When he learned about the effects of the lead on health, his mission as pastor and environmentalist became clear. He says, "The pastor is responsible for the whole man. It's very important to inform people what's around you, it's a duty. My church got involved after I got involved."

The Reverend Conley organized clinics to test the health of children who were obviously suffering the effects of elevated lead levels. At the clinics, blood tests and complete physicals were performed on children to convince the community and local authorities that something was seriously wrong. In 1984, the church filed a lawsuit against the smelter, demonstrating how many people were hobbled with only one leg, suffering from cancer, developing skin rashes, and feeling the other effects of lead poisoning. The church won and the smelter closed its doors.

"We had to keep organizing," the Reverend Conley says of the time following the shutdown of the smelting plant. "We kept holding meetings. We had to have another lawsuit to clean up the community because a lot of the kids were still sick. A year later, people were still calling me because kids were still suffering—nosebleeds, headaches, rashes, suffering. So we did another clinic."

The Reverend Conley continued to fight the city and the federal government with the help of his congregation. The Environmental Protection Agency (EPA) told him that if he could prove that kids were still getting sick from the lead, they would clean up the mess. "We had to go for another year of clinics," he explains, "and then we showed the EPA." Finally, the EPA was convinced and cleaned up the contamination left by the smelting plant.

The Reverend Conley responded to a careless attack on God's creation as well as a purposeful attack on African-Americans in the community. "Do you know what slag is?" he asked me in a conversation.

"Nope," I said, thinking it was the innocuous waste of broken rocks.

"You see, slag is the waste from batteries after they melt down," he explained. "When it cools it becomes hard and they got to dispose of it someway."

Slag is potent waste, containing lead and mercury. The reverend explained to me that the smelting plant in West Dallas took advantage of his community's poverty and dispose of this toxic junk as pavement for streets. They placed the poisoned slag under the feet of a community.

"Back in the days," the Reverend Conley remembers, "we had certain sections we could live in. We
didn't have no paved streets. They would just go
down to the hurting people and drop one of the big,
toxic slags in a muddy driveway. Them poor people
thought it was doing them a favor. They didn't know
it was dangerous."

The Reverend Conley enlisted the help of children.
He knew they represented a valuable asset to the
cause and hoped that involvement in an environmental issue through their church would help them
see, "You save a person not just so they can come to
your church and come to give money. It doesn't matter which church you're part of, it's not what your
denomination is, it's whether you can help people
live a better life.

"Young people were especially involved in this
effort." He beams. "It don't take much to motivate a
child when it's staring them in the face. So we
started marches and demonstrations, blocking off
streets. It doesn't sound Christian but you have to do
whatever it takes to get attention. It's like Christ
whipping the bad folks out of the temple. I done had
to do it about five times, but they knew we weren't
going to give up. Jesus had a Holy Ghost slap. I've
got some slap in me, too. So don't mess with me."
The Reverend laughs out loud.

"We have the responsibility of the Earth," says
Conley. "We are the salt of the Earth. Everyone
wants to be the sugar of the Earth. You see so many

sweet preachers, but we're supposed to be the salt. As pastor, you're the saving power—salt is what you use to cure meat, to put on meat to keep it from spoiling."

The Reverend Conley heard a call that directed him to repair the environment. "Christ doesn't end in the four walls," he believes. "We get confused about our responsibilities. We allowed our cities to get contaminated—dumps, falling-down homes. We are so stupid. The wind don't have no walls either."

How Much Wood Would Woody Replace if Woody Could Replace Wood?

We all know Woody Harrelson as the star of *The People vs. Larry Flynt,* or the acid-dropping, hype-craving lunatic murderer from *Natural Born Killers.* Before he was either of these things, we knew him as the simpleton bartender from *Cheers.* You may not know that Woody, like Clara and the Reverend Conley, is an environmental activist. He works on the front lines to protect the world's forests.

Woody has always shared an intense relationship with nature. In the sixth grade, he stood with arms crossed and eyes narrowed, preventing the other kids from stomping on anthills. In seventh grade he was asked to write a report on the environment and loss of species. The small assignment turned into fifty pages of what he describes as "an epic" tale of fact and opinion. Woody had discovered a passion.

While on *Cheers,* Woody became involved in organized environmental activism. He stepped in as Ted Danson's replacement for the American Oceans Campaign a few times and gradually learned that ideas are most effective when translated into action.

Woody's ideas are appealing because they are clear and simple. Why waste ancient trees making paper when you can use alternatives? In the U.S., 95 percent of our original forests have already been lost. Instead of using trees to make paper, we should be smart enough to use something else. The word *paper* itself comes from *papyrus,* an aquatic plant used by the ancient Egyptians to manufacture paper and other fibrous materials. Nearly two thousand years ago, the Chinese made paper from fibrous matter reduced to pulp in water, utilizing fibers from crops such as wheat and rice straw.

Woody used an appearance on David Letterman to talk about another alternative, industrial hemp. He tossed hemp baseball hats out to the audience and asked, "Why would we want to destroy beautiful forests for phone books?"

Hemp requires few of the dangerous pesticides and herbicides used to grow crops like cotton. While the average tree in a tree farm takes seventeen years to mature, hemp grows to its mature height of sixteen feet in about three and a half months. Tobacco farmers in Kentucky could plant hemp as a replacement for tobacco and earn roughly the same income. That's why Kentucky is leading the way in fighting for hemp legalization.

Hemp is not marijuana. Smoking a field of hemp would do little to make you high, although the smoke might implode your lungs if you were dumb enough to try it. In 1937, the Marijuana Tax Act muscled out hemp as a feasible agricultural product. Du Pont, which had developed petroleum-based products like nylon that compete with hemp, used the plant's relation to marijuana to take it out of the hands of American farmers.

Woody has been trying to help Kentucky's farmers regain the right to grow hemp. To get his point across he committed the heinous crime of planting four French industrial hemp seeds in Kentucky. Kentucky state law fails to distinguish between industrial hemp and marijuana, so he was arrested for his crime. Woody believes, "Once it's legalized in Kentucky, there will be a domino effect; it'll happen throughout the lower forty-eight, Hawaii, and Alaska."

The United States government hasn't always looked so harshly on the production and use of hemp. The USS *Constitution* used sixty tons of hemp sails and rigging. The first two drafts of the Declaration of Independence were written on hemp paper. George Washington, a hemp farmer himself, once said, "Make the most of hemp seed. Sow it everywhere."

Woody's quest doesn't end in Kentucky. On April 15, he shared some of his views with Uncle Sam.

April 15, 1996
Internal Revenue Service
Washington, D.C.

To Whom It May Concern:

I have decided to hold out a small percentage of
my income tax this year as a protest against the
way this government does business.

Perhaps being an environmentalist has
gone out of fashion, but anyone who has flown
over Washington, Oregon, or Northern California
and seen the clear-cuts and the immense
destruction to our forests cannot possibly let go
of that image. . . . It does not make sense that we
have a series of federal laws, including the
Endangered Species Act and the Wilderness Act,
designed to protect the forest, yet we still use
federal tax dollars to subsidize its destruction.

A government agency, namely the Drug
Enforcement Administration, has used federal
tax dollars to intimidate every state legislature
that has tried to legalize hemp. I fail to see the
rationale behind using our federal tax dollars to
block the development of a new, environmentally sound, sustainable, and economically prof-
itable industry while continuing to allocate billions of dollars to the industries that threaten
our very survival on the planet. We can, and we
must, work to develop those industries that provide both jobs and environmental protection. I
refuse to fall for the false dilemma of choosing
one or the other. Our tax revenues must buy us
both. Instead of giving ten thousand dollars to

the IRS, I donated it to aid the development of a domestic hemp industry.

Two hundred and twenty years ago, we embarked on an historic journey to "form a more perfect union" because we refused to submit to taxation without representation. Until my tax dollars stop going to subsidize destructive industries like timber, I cannot in good conscience continue to give. . . .

Sincerely,

Woody Harrelson

Woody is currently a partner in Hempstead, an industrial hemp products company based in California. He works with Emily Miggins and her organization, Rethink Paper, to turn his vision into reality. Woody is a continual source of inspiration for grassroots environmentalists trying to protect anthills. He takes time out of his busy schedule to talk with forest defenders, giving them bits of inspiration that keep them smiling like a thousand Buddhas. He is looking for ways to build small paper mills in the U.S. that would use cleaner pulping technologies and a combined fiber supply of recycled paper waste, crops such as industrial hemp and kenaf, and agricultural residues like wheat and rice straw. Woody is rethinking paper. He believes people could soon be drinking a cup of Maxwell House from McDonald's out of a recyclable hemp paper cup.

On the way to the Oscars in an Armani hemp

tuxedo, Woody was asked by *Entertainment Tonight,* "What are you thinking about before such a big event?"

Laughing at the silly question, he responded, "What if this limo ran on solar power?"

The Wondrous Ring

I was once asked what the best type of environmentalist is. I responded with this story:

A family once owned a wondrous ring that made the wearer the most beloved person in the world. It was a simple gold band, etched with an intricate design. Whoever wore the ring was respected as the most caring person anyone had ever encountered. The ring was passed down from generation to generation, from the father or mother to the most beloved son or daughter.

Once, the keeper of the wondrous ring had three children whom she loved equally. Each had talents and attributes that plucked at the strings of the mother's heart. One, a son, unfailingly offered his time for the service of others. The middle child, a daughter, found genuine interest in what others had to say and listened avidly to anyone who approached her. The youngest child, another girl, was a leader who tirelessly inspired others to action and organized avenues for their efforts. Each was beloved to the mother, and each was equally caring.

The mother began to worry how she would choose one child to receive the ring. She worried and worried. After mulling it over for years, she called in Temascus, the famed jeweler, to make two identical copies of the ring. Temascus lived up to his reputation in every way and produced two perfect copies of the heirloom. The mother was pleased that the rings were equally brilliant and indistinguishable from one another.

When the mother neared her death, she called her three children to her bedside one by one. Each was told that he or she was the most beloved, and each was given the "wondrous ring." Her task complete, the mother closed her eyes and peacefully slipped out of this world.

As the children wept and clasped hands at their mother's funeral, they saw the replicas on each other's hands. They realized that each of them had the "wondrous ring," and they immediately pointed to the other rings, calling their siblings frauds. The funeral turned into a playground sand-throwing match as the three children yelled and screamed. For years following, they feuded over who had the wondrous ring.

Finally, worn out by their bitter fighting, the siblings decided to speak to Worabad, the oldest man in the town. Worabad had witnessed the passing of many generations. The lines around his eyes were drawn long before words recorded history. The children trusted his wisdom in deciding who had the real ring. Worabad examined the rings, but Temascus had done his job well—the three rings were indistinguishable.

"There is only one way to tell who has the real ring,"

Worabad said. The siblings waited anxiously, all believing themselves the most beloved and owner of the genuine ring. "We can only tell who has the ring by your actions. The wearer of the true ring will be the most beloved by all. It is the most beloved who has the real ring."

Each of the siblings then set out to be the most caring and beloved. The oldest devoted every waking second to the service of others. The middle child listened intently to the concerns of others and offered her thoughtful advice. And the youngest daughter sought out even the laziest of citizens, finding a way to give meaning to the lives of them all. They were known throughout as the three most caring people the world had ever known.

After years of utilizing the best of their personalities, the three children returned to Worabad. No person they had asked could choose one as the most caring, the most beloved. They asked Worabad again, "Which of us has the wondrous ring?"

Worabad hoped they would find the answer themselves, but wished to give them peace for the rest of their days. He looked at each of them, saying, "The people have told you already. None of you is more caring or beloved than the other. Each of you has the wondrous ring, as your actions have proven."

The siblings were content that they each held the real ring and continued to feel the magic of its powers in their own ways. And through the generations, with the help of Temascus and his children, every child born received the ring and wore it with a kind and caring heart.

SECTION ◆4◆

The Fear of Being Eaten by
Your Luncheon Meat

*B*eth Gausman's only mistake was to walk by a UPS truck. She was headed to City Hall in Van Nuys, California, wearing a short-sleeved cotton sundress to beat the heat of a summer day in August 1985. When she arrived at Sylvan Street, she parked her car, crossed the street, and headed into City Hall to pull some maps for a client. On her way, she walked by a parked UPS truck and her life was changed forever.

As she walked by the truck she smelled something strange. "I clutched my throat," she remembers. "As I walked by the open door of the UPS truck, I inhaled this substance. It felt like I was burning, not horrific at first, just a burning sensation in my throat."

The truck was parked in front of a fire station on Sylvan Street. When the driver felt sick, he had called his boss who told him to drive the truck to the fire station. There it rested, awaiting attention from the hazardous materials team that was suiting up in protective gear just as Beth walked past the open truck.

Beth continued into City Hall. "I went in and all of a sudden I had to go to the rest room," she explained. "I couldn't breathe, I was starting to gasp. I couldn't even move."

While Beth blacked out in a vacant rest room, two hundred and fifty people were evacuated from City Hall. She regained some semblance of consciousness and made her way out of the rest room, pale, exhausted, and moving on rubbery legs. A police officer finally saw her and approached her as she wobbled through an abandoned corridor.

"Lady, lady, were you next to that truck?" she faintly heard through a haze. She couldn't think or respond as the police officer rushed to call an ambulance. The EMTs placed her on a stretcher and took her to the nearest hospital that had a facility for the treatment of toxically exposed patients.

"They wouldn't even take me into the hospital because I was contaminated," she recalled. "The emergency people made me shower outside. I remember the nurse taking off my clothes. I felt totally helpless, like a teeny, weak sparrow, having to wash and get all of the stuff off of me."

At the time, Beth had no idea what "the stuff" was that she scrubbed off. The hospital didn't know what the substance was either as they began to treat her. They only knew that she was delirious, that she could barely breathe, and that her skin was itching and burning. They had the UPS driver in another room, suffering from similar symptoms.

All alone in her room, strapped to an oxygen tank to breathe, Beth wondered how a routine trip to City Hall had turned into a nightmare of unknown proportions. When she was released that day in "good condition," she left the hospital in the gown they had given her because her clothes were too contaminated to salvage.

The "stuff" was hydrogen fluoride. The UPS truck carried three one-gallon containers of hydrofluoric acid, which had spilled, releasing a gas that can lead to pulmonary hemorrhaging and, as Beth would learn, a number of other health problems. The highly toxic industrial chemical is used primarily for producing aluminum, pickling stainless steel, and etching glass. Since 1963, more than six hundred deaths have been linked to hydrofluoric acid.

Beth left the hospital in a friend's car, afraid to tell her husband of her ordeal. Keeping the day's events a secret was a futile effort. "That first night I got extremely hot and then ice cold," she says of the beginning of a chronic illness. "I began to get abdominal cramping and I had these horrible headaches. I was still hardly able to breathe."

She visited her doctor the next day. After performing blood tests, he optimistically informed her that she would get better.

Her illness was just beginning. Eight long weeks on a liquid diet to soothe her burning esophagus didn't help her recover. At age thirty-five she felt like an infant, reduced to eating soft, pureed food. "It was like somebody knocked the wind out of my sails," she says of her deteriorated condition. "And I never got well. I started having more migraines. I was always taking medication. I never had lung problems before, no one in my family did. I felt like my strong body was sliced up in a Veg-O-Matic."

Beth suffered three bouts with pleurisy, an inflammation of delicate lung tissue resulting in fevers, painful respiration, coughing, and liquid accumulation in the lungs. The people around her began to see Beth as a hypochondriac. They wondered how just walking by a truck could make someone sick. She felt ostracized as she became the problem, not the victim.

The nightmare Beth lived in during the daytime invaded her sleep as well. For two years, she had a recurring, terrifying dream. Beth woke up from her nightmares only to find that her daytime fright and pain were just as horrible. The mere sight of a UPS truck stopped her in her tracks or made her pull over in her car, her body frozen in fear. "From that first day forward I've had bronchitis and asthma," Beth relates with some difficulty. "Before I walked by that truck I was a singer. I used to perform at malls and nursing homes."

When her ability to sing was snatched away, Beth thought of what she could do to continue helping those around her. She wondered, "What if an old lady was standing there with her grandchild and they got exposed? Would they have survived?" She realized that her ordeal could happen to anyone.

In the place of self-pity and fear came anger and determination. "I could not stand that I went through this, and so I decided to fight the system. I couldn't live with the chemicals left out there."

There is a cost to society's technological marvels. Beth paid for the joy of bright, glistening aluminum. Most Americans don't know how close they come to being fatally poisoned each day. The transformer on your block, a truck carrying hazardous waste, and the radiation lab at the hospital all pose a health risk to your family. It is often the seemingly harmless activities that cause the worst disasters. One of our country's most deadly toxic exposures resulted from an explosion in a chemical plant that produced the binding agent to make sugar sit on the outside of the flakes of children's cereals. Hundreds of people surrounding the factory now suffer from serious diseases. Would you be willing to have your Frosted Flakes look a little more drab in order to reduce the risks to another neighborhood?

We need to be aware of the risks so that we can make choices as a society. My father, if he knew, would never have subjected the exterminator to dangerous pesticides. Beth might never have gone to City Hall that day if she knew what was in store for her. You can't make a choice without information. The Sierra Club is working with a broad coalition of environmental groups to pass new "Community-Right-to-Know" legislation that would give you the tools to find out what risks your community faces. Corporations would

be required to make available detailed lists of the toxins that they use and release in your neighborhood.
You can react to the threats by turning on the world,
buying a cabin in Montana, and living the rest of your
days in safe solitude. Or, you can take matters into
your own hands and say loudly that it's unacceptable
to put Americans at terrible risk. That's exactly what
Beth has done.

Reflecting on her afflictions—an inability to add, an
inability to read a whole page of a book, a loss of memory—Beth knew she had a reason to fight. Her esophagus had healed but the rest of her body would never
recover. With multiple chemical sensitivities, she
became sick at the smell of dry cleaning or glues. With
encephalitis, she was unable to perform simple tasks.
A physician who examined Beth elaborates on the
effects of her brain damage, explaining, "When she
takes on stress, she starts to stammer and get foggy.
Her inability to use her mind gets her depressed. She
has a limited capacity for pressure because of the
brain damage. Many tasks that before would have
been easy for her are now insurmountable. Even
cleaning out desk drawers became impossible."

Anxious to do something to prevent similar
tragedies, Beth battled her disabilities and started the
Chemical Injury Coalition. She coined the term "chemically injured" to name the problem she faced, to give
it an identity. To enlist the help of environmental
groups, volunteers put together packets of letters from
other chemically injured individuals and information

about the effects of inhaled toxins. Beth's motto: Every citizen has to get involved. Otherwise we'll die.

After forming the coalition, she attacked the problem by pushing the California State Assembly to pass a bill limiting the transport of toxic chemicals. With the help of the coalition, she launched a lobbying campaign. Their goal was to make themselves visible as victims so no one could forget the costs of the use of toxins.

When she approached Assemblyman Tom Bane, head of the Ways and Means Committee, Beth described her situation. He joined the cause immediately, asking her to speak in front of the assembly to make a case for a protective bill. The prospect of speaking frightened her. She was unsure if her damaged body and brain would cooperate in her efforts. "I knew I needed to do it though," Beth says in retrospect. "It felt like no one else was doing it at all. I'm not a neophyte and I understand that people have to push things to get things done."

Beth and her husband traveled to Sacramento, where she spoke in defense of Bill 2705. The bill proposed to regulate the transport of inhalation hazards in the same way that explosives and radioactive materials are regulated. It required established routes for the transport of materials like hydrofluoric acid, specified requirements for equipment used to contain them, and assigned specialized personnel to accompany the chemicals. It also banned transportation of inhalation hazards by truck, forcing the chemicals onto trains.

She waited two and a half hours for her turn to speak while the assembly argued over the "rock bill," an attempt to force trucks to cover their cargoes with tarps so that loose stones wouldn't crack drivers' windshields on the highway. The bill she hoped to pass was more controversial. Listening to the proceedings over what seemed an obvious issue increased her anxiety. "It was unreal, and it all seemed like a joke. But I wasn't laughing."

When she finally walked up to the podium to tell her story, Beth's body cooperated halfheartedly. "I got really raspy, really hoarse," she explains. "I guess it made a difference, though, because everyone stopped and listened. You could hear a pin drop. The whole place was focused, everything went quiet." She went on to tell her story to the captivated audience, explaining her illnesses and her innocence. She finished saying, "I was hurt. I was a pedestrian. I don't work with chemicals, I don't handle chemicals, and I don't work a high-risk job. I was exposed just crossing the street."

"It took a year of campaigning," she says of her fight to pass the bill. "I learned how big the problem was and I wanted to make a difference. I spent my own money, and it took a year, but we won. Now one hundred and ninety toxics have to go by train."

Beth operates a business called Safety First, which is creating "Every Parent's Handbook to Toxics," a guide to help parents protect their children. Slowed but undaunted by the effects of the toxic air she inhaled, Beth fights for better protection of our environment and our health. Inspired by her own encounter with a challenge, she took the first step in ensuring that chemicals don't injure the health of the innocent and unaware. Having her health stolen from her made her overcome her fears. Despite her numerous disabilities, she prevailed.

Beth never did anything to deserve her ongoing ordeal. Her exposure was random. As Beth knows, we are playing a game of Russian roulette with ourselves. Will the next UPS truck you walk by poison you? Probably not, because of the efforts of Beth Gausman. Will the tap water make you sick? Will your next hamburger send you to the hospital? Will you discover that your house was built on a toxic landfill? You can spend time wondering, or you can get to work to make sure that your grandchildren won't have to ask those same questions. It's your right to know. Demand it.

Train Spill in the Sierra

June Spalding lived with her family in the small town of Dunsmuir, California, population 2,400. On a summer night her oldest son was out partying with some friends near the Cantara Loop, an area close to the Sacramento River, near Dunsmuir. He said that during the night he heard a *Bam!* followed by a long silence. He and his friends ignored the loud crash because train derailments are nothing unusual in the area.

Little did they know that this derailment was unusual, even for the Cantara Loop. Fourteen thousand gallons of metam sodium, a highly concentrated herbicide known as Vapam, spilled into the Sacramento River. It poisoned the wildlife, the water, and the air that the citizens of Dunsmuir breathed.

Metam sodium is an agricultural herbicide used on carrots, cotton, melons, and tomatoes in the state of California. This herbicide, one that is sprayed on our food, is poisonous if ingested and irritates the skin and other mucous membranes. That's before it's mixed with water.

When mixed with water, metam sodium decomposes, releasing highly toxic fumes of nitrogen monoxide. One of the breakdown products of the chemical is a relative of the mustard gas used by Iraqi soldiers in fighting Kurdish rebels. Vapam itself was also once considered for use as a war gas. The gas is described as a "poison by inhalation, skin contact, subcutaneous and intravenous routes. . . . [It is] a military blistering gas. [It] strongly affects the skin, eyes, lungs, and gastric system. Pulmonary lesions are often fatal. It penetrates the skin deeply to injure blood vessels. Minute amounts can cause inflammation. Secondary infections are common." Our society should not be using lethal weapons of war to ensure that our fruit is shiny.

While the spill of metam sodium unleashed all of these health hazards on local communities, the list of actual effects felt by citizens was even longer. Dr. Cindy Watson, a Los Angeles physician who was called in to examine patients affected by the poisonous spill, was shocked by what she encountered. Loss of color vision, tremors, damage to the nervous system, headaches, breathing disorders, sinus infections, increased sensitivity to chemicals, sexual dysfunction, high blood pressure, extreme fever, and loss of memory were just a few of the symptoms she encountered in treating victims of the spill.

Dr. Watson's experience in toxicology led legal counsel to call her in to examine the patients in Dunsmuir. "I had been involved with kids who had suffered from chronic health problems that drugs didn't seem to cure," she said. "It was often the environment that was making them sick—lead poisoning or some other chemical. I had seen patients for pesticide poisoning and I knew how to take care of someone who had been made toxic. And I knew what difficulties they face." Dr. Watson is part of one of the fastest growing specialties in medicine—environmental medicine. These doctors help patients who suffer from the toxic overload of the twenty-first century.

Watson examined prisoners from the Crystal Creek Detention Facility in Shasta County who were brought in to remove the dead and dying fish and wildlife that floated in the river's waters and spotted its shores. Prisoners from Crystal Creek who accepted the voluntary job of restoring the river were given latex gloves, hip waders, and painters' masks to protect their bodies from the toxic chemicals.

Southern Pacific, the company that owned the train, knew how toxic the spill was. The workers from Southern Pacific wore specialized suits during the cleanup and knew the chemical was a health threat. None of the prisoners enlisted to help knew until their health began to falter.

Just days after exposure to metam sodium, the prisoners began to suffer from serious headaches and nausea. Then red rashes and blisters developed on their feet and ankles. "Many of the prisoners, all of whom were about twenty to twenty-four years of age," explains Watson, "were experiencing sexual dysfunction and having difficulty maintaining erections. . . . Many had elevated liver function and diarrhea. They had difficulty sleeping and suffered irritability."

The prisoners' sentences were extended by the carelessness that led to the spill. For some, it will become a life sentence.

"There were a large number of miscarriages and suicides afterwards because of toxic effects on the brain," lamented Dr. Watson. Many victims now face multiple chemical sensitivities. "When the liver, which processes chemicals in the body, is full, the immune system can't handle more chemicals because of the original injury. Artists who are overexposed to turpentine can become chemically sensitive. These people are now chemically sensitive. The people up there were angry and felt helpless."

Wayne Cunningham feels helpless. The metam sodium ravaged his body. Months after the spill he complains, "My eyesight is going away and I can't remember things. I have massive headaches, loss of eyesight, and loss of memory. I get irritable. You can't function right. I know numbers really well. I had to look up my mom and dad's telephone the other day."

Wayne is a mechanic who maintains and operates heavy machinery. He was sent in to manage a barge under the direction of Southern Pacific. He reports, "There were no warnings that we should expect anything harmful to happen. There was no 'This will do that to you and that will do this to you.' Their thing was that if we contacted any of the chemicals through the meters measuring for its presence, we were to fire a red flare that meant get off the lake, get in a boat and take off."

No red flares were fired to warn Wayne that he would collapse after one day of working on the barge. He had a fever of 105 degrees alternating with hypothermia, and skin rashes covered his body. He was one of hundreds taken to the hospital after exposure to the herbicide and its by-products.

Wayne remembers four or five newscasters on the hillside when he collapsed, none of whom filmed him even as he was carried out on a gurney with an IV in his arm. They hadn't made the connection.

"Three weeks after the spill, people were told it was all right to camp along the river," says Dr. Watson, describing a family who had done just that before becoming her patients. "They've all developed persistent conjunctivitis. . . . They have developed neurological problems. The children have blurred vision, tremors, fatigue, headaches, skin rashes, blisters."

Wayne Cunningham now wonders about his future. "I just want to know if this stuff is going to knock ten or fifteen years off my life. I have twelve grandkids, and I don't want to leave them before my time."

Dr. Watson remains frustrated with the attitude of carelessness and incompetence that characterized the incident. Her medical opinion is, "These things should never be used—not even manufactured, let alone used."

She remains dedicated to meeting the challenge. "We've got to do something. I'm doing something in my own little way. I see the front lines of the problems. These patients are the canaries in the coal mines."

Lois Gibbs knows all about being a canary in the coal mine. After helping her community in Love Canal, New York, move off the landfill that they had unknowingly been sited on, she went on to found the Citizens Clearinghouse on Hazardous Waste (CCHW). CCHW is the preeminent grassroots organization helping neighborhoods across the country fight the threat of toxics. Lois says that we have to move from a mindset of "NIMBY" (not in my backyard) to "NIABY" (not in anybody's backyard). If a substance is too unsafe to be near your children, it's unsafe to be around anyone's children. If a chemical can't be used safely, we should question whether it should be used at all.

Although pleased with her successes, Beth Gausman would agree. After telling me her story, she said, "I did put all of the toxics on trains. But we can't just move the stuff to a safer form of transportation. We have to stop making it altogether." Citizens of Dunsmuir, California, agree as well.

Job of Politics: The Portrait of a Seamstress in a Straitjacket Factory

The residents of Dunsmuir know that eliminating the production of the most dangerous toxics altogether is the only way to prevent spills and life-threatening accidents. The political process is the likely forum to protect them from the dangers of hazardous materials. But with the amount of money poured into the wallets of members of Congress by the same corporations posing the risks to society, their fight is an uphill battle.

I keep a very select group of members of Congress on a special mailing list in my office. We call it the globe list. You need to have special qualifications to get on the globe list, and I send a personally auto-graphed postcard to every member of Congress who makes it. On one side of the postcard is a satellite photograph of the Earth taken from space. On the other side I write, "Wish You Were Here," and draw a little happy face. I'm sure you'll agree that Representative Tom DeLay (R-Texas) belongs right at the head of the globe list.

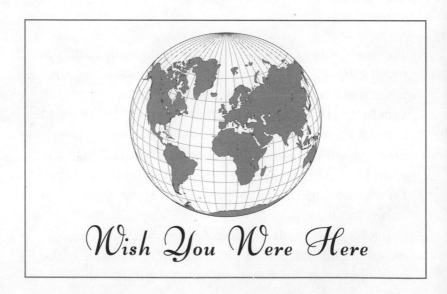

When Republicans took control of Congress in 1994, DeLay, a former exterminator from Houston, saw an opportunity to promote his wacky world view. DeLay began his political career in the Texas legislature in 1978. As an exterminator, he opposed all federal safety rules, like requiring workers to wear protective hard hats while tunneling under houses to attack termites. Pesticide regulations, established by the EPA to protect workers and the general public, "drove him crazy." He quickly earned the nickname Mr. DeReg.

DeLay has said that he can not name one federal regulation that makes sense. He sounds like my good friend Rush Limbaugh.

Soon after he arrived in Washington, DeLay began pushing for the end of government health and safety regulations. To get the ball rolling, he gathered a powerful and well-financed group of lobbyists under the name of Project Relief. With a name like that, DeLay's pack of lobbyists sound like a support group for Mother Teresa. That's exactly what this special interest group would have us believe.

The membership of Project Relief includes people from the oil, tobacco, pharmaceutical, and utility industries. With a gaggle of Gucci-wearing lobbyists and a half-million-dollar communications budget, Project Relief went to work for Tom Delay by funding campaigns of other antienvironment members of Congress and by drafting legislation that would cripple the U.S. Environmental Protection Agency.

DeLay began his campaign to halt regulatory laws by writing a letter to the Clinton administration requesting a two-hundred-day halt on federal rule making. The request was unequivocally refused two days later. DeLay then turned to legislation to accomplish his objectives.

He enlisted Gordon Gooch, a rotund Texan who lobbied for petrochemical and energy interests, to draft the legislation. The legislation sought to stall upcoming action on regulations that would place tougher rules on meat inspection to prevent *E. coli* infections and tighten labor standards to reduce workplace injuries. Before Delay's halting legislation hit the House floor, Project Relief lobbyists warned lawmakers that their votes would be considered in future campaign contribution decisions. As debate raged over the legislation, those same lobbyists fed representatives the arguments they needed to defend the indefensible.

In the end, the moratorium on federal rule making passed the House with a vote of 276 to 146. DeLay and Project Relief successfully convinced and coerced enough representatives to think of campaign contributions over public safety.

That's exactly why the number one challenge to our environment is campaign finance reform. Almost every member of Congress who supports drilling for oil in the Arctic National Wildlife Refuge receives contributions from Exxon. Almost every member of Congress who votes against higher fuel efficiency standards received money from the automobile and oil industry. Is there a connection?

From 1993 to 1996, anti-environment special-interest groups gave more than $34 million in congressional campaign contributions. By opening their wallets wide, corporations attacked the quality of our water, supported clear-cutting forests, slashed at public protection programs, and wrestled for the power of the Environmental Protection Agency.

Rather than look at individual programs and laws, Congress, backed by special interests like Project Relief, looked to pass a regulatory lobotomy. Proposed legislation could have weakened every public health and environmental safeguard enacted in the last twenty-five years.

The law that crept into Congress under the guise of "regulatory reform" had dubious origins. Most people believe that representatives and senators, or at least their aides, write the laws on which they later vote. Not so in the case of HR 9, the "regulatory reform" bill considered by Congress.

Oil and chemical companies worked hand in hand with high-powered law firms to create the legislation that claimed to streamline government for the benefit of America's citizens. The legislation required in-depth studies to determine industry cleanup costs in comparison to the much less tangible societal benefits involved in any new public health and environmental protection measures. If the industry costs exceeded the benefits to humans at all, standards would not be implemented or enforced. But how do you quantify the cost of cancer in a child?

Although the "regulatory reform" bill passed in the House, its companion Senate bill was halted by senators with an eye on our environment.

The Joy of Politics

From toxics to train spills to campaign finance reform, America faces many challenges. Anyone with access to a television, newspaper, or computer can see the signs. Riots fueled by racial tensions and economic inequalities break out in our cities. Less than 35 percent of eighteen-to-twenty-four-year-olds voted in the 1996 election, the lowest percentage since eighteen-year-olds were given the right to vote. Those who did vote were responsible for electing Sonny Bono to office.

Our elections have turned into a glorified version of *Star Search*. Clinton performed his saxophone act on *Arsenio*. Former President George Bush jumped out of an airplane and he's not even running for office; he just misses the show.

Dan Rather, Peter Jennings, and Tom Brokaw serve as our daily reminder that dishonesty, murder, kidnapping, and abuse characterize much of our country, from the most likely perpetrators to the supreme courts of law. We reelect felons shortly after their release from jail. Marion Barry's campaign slogan, "A Chicken in Every Pot," became "Pot for Every Chick," and he managed to sail right back into office.

Before Marion Barry became the most famous mayor to be elected as a convicted felon, Mayor Buddy Cianci of Providence held the title. Buddy went to jail for beating his wife's lover and burning him with a cigarette while the cops held the man down. When he got out of prison, he won the race for mayor in a landslide election. His campaign slogan was, "I want to make love to Providence. Providence hasn't been made love to for too long!"

More people are incarcerated today in America than in any other society in the history of the world, which may explain why we can't fill our public offices with people who haven't been to jail. If that doesn't frighten you, did you know that Rush Limbaugh is now one of the leading sellers of ties in America? That's a crisis. It's bad enough to listen to him spout off. I don't want to have him around my neck. I blame him for taking some of the joy out of politics.

Rush once defied his viewers to name one government program that has worked.

I can name a few:

Child labor laws keep kids in school rather than in sweatshops.

Safe drinking water protections safeguard Americans from disease.

Because of the Endangered Species Act, the bald eagle, our national symbol of freedom, is no longer endangered.

The Superfund Program, a multibillion dollar effort, has led to the cleanup of hundreds of hazardous waste sites.

Since 1973 we have phased out the use of DDT, a deadly carcinogen once widely used on our food.

America's National Park System is the envy of the world, incorporating 374 units from the towering ice sculptures of Glacier National Park to the rolling dunes of the Cape Cod National Seashore and the murky, mysterious swamps of the Everglades.

The environmental successes of the past can be a source of hope for the future of democracy. If we can save the bald eagle and a river that once caught fire, then we're ready to take on any challenge.

It's probably no use to try to convince Rush, but I can't forgive him for setting a tone for the debate that turns America off from the belief that the political process can help.

Rush is not a "good news" type of guy. He only seems happy when politicians are duking it out and there's blood on the floor. Constantly cynical voices chase Americans away from the political system. Although it's not perfect, it has had some incredible successes.

My friend Jessica Tully is the field director for Rock the Vote. When she tries to register people to vote the most common excuse she faces is, "Politics is a waste of my time. It doesn't do anything." As much as I'll bash some members of Congress, I respect them for investing themselves in the project of democracy. They believe that their service will do something. But we fail to celebrate our past successes to make the rest of America believe.

The environment can serve as the issue that ignites a newborn faith in politics.

People say that regulations reel out of control, but they think it's important for the government to clean up an oil spill. They can complain about wasted tax dollars, but not without being one of the millions who visit our government-funded national parks each year. They can slight the public education their children receive, but not without acknowledging that their kids learn more about ecology than they learned. We need to expose this confusion and use it to move forward.

The environment offers us a chance to reinstill hope because it is an issue on which everyone agrees. Everyone wants clean air to breathe, safe water to drink, and poison-free food to eat. Rallying around the environment is not a question of being green—it's a question of whether you're red, white, and blue. If we don't have a world to live in, you can kiss all of your other concerns good-bye.

With a newborn faith in politics, we can also bring the joy back. Representative Jim Maloney (D-Connecticut) knows what I mean. I went to Connecticut to help him get elected.

I arrived on a gray day in Shelton to announce the
Sierra Club's endorsement of Jim for Congress.
Shelton is a town going through some hard times.
When manufacturing plants closed in the 1940s they
left behind a toxic legacy that has kept new busi-
nesses at bay.

Jim, a big fellow with a toothy smile and a firm
handshake, looked uncomfortable in the poorly tai-
lored new suit he wore for the event. He was trying
to look congressional. Polls showed that although 98
percent of likely voters knew his opponent, only 35
percent could pick him out in a crowd.

As a member of the state legislature, Jim had
worked tirelessly to reclaim Connecticut's future by
dealing with its toxic past. He had fought to clean up
a toxic waste dump and create an urban park in its
place. His opponent in the congressional race was a
Gingrich clone. Jim was in for a tough race.

I did a few radio interviews and then drove down
to a Superfund site where his press secretary had
planned a press conference. Jim had charts, talking
points, hair gel, and a local citizen.

One problem. No press showed up.

Without missing a beat Jim asked me to come with
him to the offices of the local newspaper. "If they
won't come to us," he declared, "we'll go to them."

I was enjoying myself. Jim Maloney has an infec-
tious exuberance. He was determined to bring his
ideals to Congress.

We arrived at the newspaper office and Jim

vaulted in. The reporters glanced up, and then quickly returned to their crossword puzzles. They had no idea who he was. This didn't stop Jim. He grabbed his charts and started walking toward one reporter when . . . *SLAM!!!* He tripped over a phone cord and stumbled to the floor, bringing every phone in the office cascading down on top of him.

Now he had their attention.

He quickly picked up the phones and dusted himself off. Without missing a beat, he began his pitch to a reporter who looked more frightened than amused. Jim kept pushing, eked out a victory, and today serves in the House of Representatives.

That's the joy of politics. Democracy gives anyone a chance to throw a hat in the ring to help steer the direction of the country. We need to celebrate the system, with all of its foibles. If people knew how funny, fallible, and human our elected representatives are, they'd probably get more involved. I find better comedy on the floor of the Senate than in the last three seasons of *Saturday Night Live* combined.

Don't Boot on Newt

Political parties are sometimes like network TV. They don't give a damn what they're saying or selling as long as it increases the ratings. Politicians often run for office on the weaknesses of their opponents rather than on their own merits. President

Reagan tested a commercial using the image of cop-
ulating rhinos to predict the effects of Mondale on
the American public. President Bush exploited deep-
seated racial divisions to attack Michael Dukakis on
the prison furloughs that freed Willie Horton, who
raped and killed again. President Clinton exploited
the weaknesses of age to beat Dole.

A detailed 1995 memo written by Representative
John Boehner, entitled "Think Globally, Act Locally:
A Pro-Active, Pro-Environment Agenda for House
Republicans," proposed public relations stunts that
would cover lawmakers' actions in attacking our
environment. The report encouraged deception, saying, "You must engage this agenda before your opponents can label your efforts craven, election-year
gimmicks. Remember, as a famous frog once said, it
ain't easy being green."

The memo suggests that anti-environmental members of Congress hold a beach cleanup in their neighborhoods to look "green" to their constituents. The
memo went on to suggest tree-plantings, recycling
events, and other "excellent earned media events to
counter the environmental lobby and their friends in
the eco-terrorist underworld." Nonetheless, I decided
to go to the 1996 Republican Convention in San Diego
to celebrate the accomplishments of past Republicans.
I went thinking of the photo of Republican President
Teddy Roosevelt standing with John Muir in the newly
protected Yosemite Park. I hoped to begin talks with
Republican leaders that would lead to greater cooper-

ation between environmentalists and Republicans, but I should have known they wouldn't roll out the red carpet for me. I felt like Jeffrey Dahmer at a family picnic.

When I arrived in San Diego, I found that there was no chance in hell that Speaker Newt Gingrich would meet with me. Instead, he visited the San Diego Zoo. During Newt's day at the zoo, he posed for photo opportunities, shook hands with the little children out for a day of enjoying wildlife, and projected an image of caring and concern. The visit was a smokescreen for his efforts to weaken the Endangered Species Act.

Most people fell for his tactics. Newspaper photographers snapped pictures of his magnanimous grin next to the faces of the less fortunate species of the world. Not all of the animals were fooled, though. It was reported to me that later when Newt approached a tiger, the large cat turned to look at him, eyes scanning his large frame as if it were sizing up a Thanksgiving turkey, paused, and then proceeded to throw up on him. You can't fool the animals.

Rather than welcoming my participation in their event, the Republicans quartered me off as a protester. Despite my peaceful agenda, I spent most of my time with the other "protesters" in a fenced-off half-acre plot a half mile away from the action. People who have clean bedrooms don't pull the shades and lock the doors when Mom shows up for a surprise visit.

Newt, if you let me into the convention next time, I promise I won't throw up on you.

Steven Seagal Meets Bob Weir in the Battle Royal

I had never been to Kentucky before I became president of the Sierra Club, but my goal was to visit all of the Sierra Club's sixty-five chapters. It was time for a visit to Kentucky, and all I knew about Kentucky I had learned from *The Beverly Hillbillies*.

Kentucky has lost almost 80 percent of its wetlands to development. Family farms have been subdivided into housing complexes. Cement foundations now fill the floodplains. As the wetlands disappear, farms and new developments face an increased threat of flooding.

I am often asked what the big fuss about wetlands is. Imagine wetlands as one hundred large sponges covering a dinner table. When you pour a bucket of water on the table, the sponges absorb it all, saving the floor from being doused with water. Now remove eighty of the hundred sponges and pour another bucket of water over the table. The water splashes off, soaking the floor around the table and destroying the rug in your dining room.

In late 1996, I spoke to empty auditoriums and rooms speckled with glassy-eyed Kentucky residents. I even used my favorite wetlands line from Brett Hulsey, Midwest regional representative of the

Sierra Club: "Funny thing about floodplains, they just plain flood." I had come to Kentucky to discuss the loss of wetlands, but no one seemed to give a damn. Although I drew a few die-hard environmentalists to my speeches, their interest lay more in meeting the weird kid who was the president of the Sierra Club than in discussing the crisis of wetlands destruction.

In January 1997, Kentucky was hit by the most destructive flooding in its history. The Ohio River overflowed its banks, sending water into the streets of towns and the basements of houses. Farmlands were swamped and cars floated up into trees where they rested after the water receded. The Confederate Acres, a housing development in the floodplain, was completely destroyed. During my first visit, I learned that the owner of Confederate Acres had chased concerned citizens away with a gun when they came to protest the development's expansion into more wetlands.

I made a second trip to Kentucky in 1997 to talk to people about the flood damage. This time I was more popular than Jennifer Aniston's haircut and had trouble balancing my speaking schedule with TV and newspaper interviews. I spoke to rooms packed with people eager to hear how the preservation of wetlands controls flooding.

Lesson learned. It's easier to sell something that people are buying than to convince them to buy something that you're selling.

Before I left for Kentucky for my second trip, I
arranged some backup appointments in case I
encountered the same reception as the first trip. I
scheduled a trip to Atlanta to meet with Home Depot
to discuss a boycott on old-growth redwoods.

A few months earlier, September rolled over the
Headwaters Forest in northern California and the
heat of the summer began its journey to other parts
of the world. Misty clouds, moisture heated and suspended in the serpentine halls of the old-growth forest, retreated out to sea. The silence of the forest
was broken by a booming chant.

"Save Headwaters Forest now!" The cry rang out
just outside the gates of the Pacific Lumber Company
in Carlotta, California, on the morning of September
15, 1996.

I had driven seven hours to address a crowd of
more than two thousand people who were outraged
at the recklessness of a greedy corporation. We were
protesting the Pacific Lumber Company's plan to
begin salvage logging in the ancient redwood groves
it owns.

I spoke into a sound system powered by a large
solar panel and a generator powered by the pedaling
feet of my companions. I shouted through the dissipat-
ing fog of a warming morning, "Six ancient groves,
sixty thousand acres. Save Headwaters Forest now!"
The cry came back at me with the strength of thousands, and we played Ping-Pong with it until our
throats went dry. With a high five, Bonnie Raitt took

the stage and rocked the crowd with a song.

We came to support the work of tireless activists like Judi Bari of Earth First! Judi and others dedicated their lives to the protection of this ancient forest, sometimes chaining themselves to the trees slated for removal. It was Judi's last large rally before her death from breast cancer.

The Pacific Lumber Company intended to begin salvage logging in three of six groves that comprise the largest unprotected tract of ancient coastal redwoods in the world. Only 4 percent of the original old-growth redwood forest remains.

All told, the ancient groves cover 7,500 acres, and a surrounding buffer zone of second-growth forest covers an additional 52,500 acres. Together the trees are a biome, an integrated system of life that supports species like the endangered Coho salmon, which spawns in the watershed of the forest. The system provides a safe passage for wildlife between Redwood National Park and Humboldt Redwoods State Park. It relies on all of its parts for sustainable living.

Salvage logging signals the end of the forest—death by attrition. Salvage logging steals away dead trees, the woody material that decays into the soil, committing grand larceny of essential forest nutrients and diminishing the prospects for young trees. The "dead" trees are so full of life that if you take a piece of bark from a redwood tree and stick it in a cup of water, a new tree will sprout. But if you take away these nutrients from the forest, nothing will grow.

Before I went to the rally, Ed Wayburn, now ninety years old, the hero of the debate over the Sierra Student Coalition and a legend for saving large parks in Alaska, sat me down in his office.

"Adam," he said, shaking his head, "I kick myself when I think that Headwaters could be lost. When we were forming the plan for Redwood National Park, we intentionally passed over the lands owned by Pacific Lumber." He paused to stare out the window at a police car that wailed down the street and past the office. "Pacific Lumber was a family-owned business. For a hundred years they had managed their forests well, never threatening to cut the ancient trees. But then the 1980s came along, greed took over, and Hurwitz decided that trees were just another resource to be liquidated."

Houston financier Charles Hurwitz acquired Pacific Lumber in a 1985 hostile takeover. He had previously taken over Kaiser Aluminum. Their waste emissions became the worst in the industry within five years of his taking control. In the case of Pacific Lumber, Hurwitz's takeover resulted in a tripling of logging activity in the buffer zone surrounding the ancient redwood groves.

And why were these trees threatened? These markers of time, these records of history would find their way into decks, patio furniture, and decorative borders for hot tubs. It was time to take the campaign to the people who use redwood, to show them the costs of their patio furniture. The environmental community came together for a boycott. We held a press conference in L.A. to announce the campaign.

We asked Steven Seagal, the ultimate action hero, to kick off the event. We figured he would blast Hurwitz with karate chop anger. He swaggered up to the microphone and quietly asked, "Isn't it strange that every time something goes wrong in this world it seems a big corporation is behind it?"

He stepped off the platform and looked at me asking, "Was that okay?" I gave him a thumbs-up, surprised yet pleased at the style of his statement.

Then Bob Weir, former guitarist for the Grateful Dead, came to the stage. He brought with him a more spiritual argument and threatened of oncoming ice ages if we didn't protect our forests. Then he read a passionate open letter to Charles Hurwitz:

> Maybe thirty years ago, I was on one of my first band tours. We were in the Pacific Northwest, between somewhere in Washington and some other where in Oregon. The road took us up on a ridge, from which we could see around us for many miles in all directions. To the west we could see a weather front moving high clouds in from the Pacific. To the north and south, where the front came parallel with us, we could see a mist rising up from the forested foothills all around us. And when this mist joined up with and seeded the clouds passing overhead, it turned to rain and snow which then

fell on the mountains to our east. Scientists call this regular phenomenon evapo-transpiration. I wish you could have seen it.

It was breathtaking to behold, but as we watched, we had a firm realization that we were witnessing something even more beautiful than our eyes could ever take in. We saw how the rain falls to the earth, and there rises the grandest protein of all life—the forest, awesome in its size and complexity. The forest, in turn, holds this moisture until the next storm front comes through when again the mist will rise, the clouds will seed, and rain will fall. Life causes life. Heaven and earth dance in this way endlessly, and their child is the forest. . . .

And so there we were, experiencing an epiphany, watching that grandest and most glorious dance of life—of which we are just a tiny part—awed by a magnificence without beginning, without end. . . .

A couple of years later, we were making the same trip and we came to the same place, but the forest was gone; now the land was bare. The same weather patterns moved through, but now no mist rose up to seed the clouds and the rain no longer fell so much on the mountains to the east. I was still very young, but it seemed altogether wrong to me that we should destroy something so big, so far beyond our understanding. What unimaginable arrogance!

Perhaps you should go and sit for a while in one of your clear-cuts and think this over as you listen to the desolate sound of the wind as it blusters unhindered past your ears, bereft of the trees that used to tame it. Then, go and

spend some time in the magnificence of the
ancient forest you plan to destroy and perhaps
you will hear a voice much older, wiser, deeper,
and gentler than ours; it's there. And if you do,
then when your time comes, you will die a far
richer man than you might ever have imagined.

The boycott then began; we took on Home Depot,
which sells 10 percent of all consumer lumber in the
United States. Home Depot is the big daddy. Mark
Eisen, a squat man with a busy mustache floating
above his mouth, is the director of environmental
affairs at Home Depot.

I called him to explain: "Look, Mark. We need
Home Depot to come along on this." We knew that
support from a major supplier would not only slow
the flow of redwood but also garner more media
attention. I always assume the best about people, so
I went after his morals first. "You need to do the
right thing. The bottom line is that you've got a big
orange sign, and if you don't sign on to the boycott,
all of our activists will head right toward that orange
bull's-eye."

He replied with a string of justifications about the
challenges of a growth company, plans for doubling
the number of orange signs, and trying to make it all
work for Home Depot. I told him that his company
faced an imminent and public battle and he'd better
contemplate the merits of stocking old-growth redwood.

Home Depot didn't take the bait. So we turned to

their competitors, Orchard Supply Hardware and Home Base. They were pleased to find they had an opportunity for a competitive advantage over Home Depot and welcomed the chance to show the world that Home Depot had no concern except for the bottom line.

Two weeks later, I received a fax from Home Depot's public affairs department. They signed on to the pledge. Bingo—three major allies with one piece of bait.

If there's no market for old-growth redwoods, we won't have to worry about the forces of greed cutting them down. Redwood is just the beginning. As we develop better tools for the identification of the sources of lumber, we will expand the boycott campaign to reach all types of ancient-forest timber. Each day we find a new secret use of old-growth timber. Have you had a glass of orange juice recently? In some brands, the suspension agent that ensures that the pulp doesn't settle at the bottom of your glass is sawdust from old-growth trees.

It is unacceptable for corporations concerned with short-term profit to rape our forests—our children's natural inheritance. I want my kids to be able to experience the same wonders of the cathedral forests that I have been fortunate enough to see. It is humanity that would destroy them . . . and our future.

The Man Who Wanted to Be the Sun

Once, in the magical land of Phunk, there lived an old stonecutter named Bootsy.

Bootsy woke up each morning just as the sun peered over the trees and pushed his wheelbarrow many long miles to the foot of the mountain.

The sun beat down mercilessly on Bootsy, blistering the skin on his neck. As he walked and walked he would look up and curse the sun.

When Bootsy reached the base of the mountain, he would lie down the wheelbarrow, take out his little hammer, and *Clink Clink Clin-C-Clink,* he would pound out cobblestones. All day, the sun beat down on Bootsy as he worked, prickling his ebony skin.

With the wheelbarrow full Bootsy began his trek toward home, all the while struggling under the weight of the sun. Sometimes as he made his way home the clouds would come out to cover the sun. Relieved, he would begin to think of slapping out grooves on his bass. When the clouds opened up and sent their droplets flying to the ground, the rain splashed sweaty salty rivulets into Bootsy's eyes and soaked him with extra weight. He would plod on, now cursing the clouds that had sheltered him from the sun.

One day, just as he reached the bottom of the mountain, Bootsy lost his groove, slipped, and cracked his shin on his wheelbarrow. The sun beat down particularly hard that day, and in a fit of delirium, Bootsy began to rant and rave.

From up above him, he heard a deep and bellowing voice ask, "Who dat?"

He looked up and saw a huge grin on the face of the sun. Bootsy had heard tales of those who had spoken to the sun and asked it for favors, but they were late-night stories after jam sessions and a few too many sips of Uncle Jam's potions.

Bootsy's mind reeled. He began to think that the stories of the magical powers of Phunk were true. And in a blind rush he blurted out, "I wish I were the sun. I wish I had the power of the sun, and then I would be truly powerful!"

Just as he finished his words, he felt a strange burning sensation throughout his whole body and before he knew what had happened, he was the sun.

He rejoiced. He thought, "Now I am powerful. Now I am the sun and I will beat down on the backs of the Phunkers, making their skin prickly with just a glance from my eye. And I will light fires in the forest and chase animals into the shade of the thickets."

And for many days, Bootsy the sun went about being the most brilliant and powerful sun that the land of Phunk ever felt. And he was happy. One day, as the sun traveled through Zigaboo Valley, a little dirty cotton ball of a cloud drifted over the horizon. The sun paid it no heed.

Slowly, another little ball of cloud floated up, and another and another. Gradually, the little clouds joined hands until they became one big cloud. The big gray cloud blanketed the valley from Bootsy's rays. He shone as hard as he could but he couldn't get through. Bootsy started to curse once again.

"I thought I was powerful," he fumed. "I can burn people and start fires. I can drive the strongest of men to their knees. But now, I am not powerful. These little clouds, they are truly powerful because they can stop the rays of the sun. If I can't shine my rays to the earth I have no power at all." The sun paused to try to poke through the clouds one more time, but he was left tired and cranky.

In despair, the sun cried out, "I wish I were the clouds. I wish I had the power to block the rays of the sun." No sooner had he said it than he felt a cool moist wave envelop him and suddenly Bootsy was the clouds.

"Now I am truly powerful," Bootsy said as he floated on the breeze, sprinkled water to make the crops grow, flooded villages by feeding rivers, and blocked the rays of the sun. "I can make life grow. I am as powerful as anything could be!" Bootsy spent many days on the wings of the wind. And he was happy.

But before long the clouds caught on something. Bootsy tugged at his misty form but couldn't get back on his windy transport. He looked down and saw a mountain holding him back. He rained down on the mountain, trying to wash it away. But Bootsy's teary droplets fell in vain. The mountain was too big and too strong, and the clouds began to curse.

"Where is my power now?" lamented Bootsy. "I

thought I was powerful because I could flood the rivers and feed the trees. I could stop the rays of the powerful sun. But this mountain is truly powerful because it can stop the clouds that can stop the rays of the sun."

And with a thundering clap, the clouds shouted, "I wish I were the mountain so that I could be all-powerful and stop even the clouds."

Suddenly, Bootsy melted down and felt himself becoming solid. Before he knew it, he was the mountain, grasping the clouds that could not struggle away. And the mountain, formerly the clouds, formerly the sun, formerly a man, thought, "Now I am powerful. I don't have to push a wheelbarrow or pull myself over the horizon or ride on the back of the wind. I can just sit here. The sun cannot melt me and the clouds cannot wash me away." And Bootsy, the mountain, was happy.

Bootsy spent a whole weekend gloating over his new position. He watched the colorful countryside of Phunk being scalded by the sun and washed by the clouds, and he knew that he was immovable.

On Monday morning, the mountain felt a little pain in his foot. A sharp pinprick shot up his side, and he heard a noise from far below. *Clink. Clink. Clin-C-Clin-Clin-C-Clink.* He looked down to see a little old man in a green felt hat. The man's neck was prickling in the heat of the sun and he cursed as he slowly removed piece after piece from the mountain. And as the wheelbarrow filled with stones, Bootsy realized that even the mountain could be taken down.

The mountain sighed, seeing that real power lies in

the power of one. One person, moving one step at a time, can conquer anything.

And as Bootsy's memories rushed in, he listened to the *clink* of the hammer and said, "I wish I were just a little old man."

SECTION ◇5◇

Who Put the Twinkies
in My Canyon?

Fiddling Around

I woke up with the light of morning, planning to hike Orange Mountain in the Green Mountain National Forest. The view from the peak sweeps the valley below. I was prepared for eight miles round trip, five thousand feet split between climbing and descending; the peak sat well within reach. I was eager to conquer the mountain. I checked my gear, my supplies.

Picking myself up and out of my sleeping bag, I took my plastic canopy down from its sapling posts and tried to rest the ache from my back after a night on a granite mattress. I reached down to touch my toes.

As I bent down to the ground, a delicate fiddlehead broke up through the soil, reaching to find the spring. Its radiant green bud made a gaudy accessory on the spongy brown soil of the mountain. The fern looked unnatural, like a neon marker placed there by a hiker who preceded me. At first, I thought it was a bit of plastic or a candy wrapper. As I looked more closely, I saw the spindly veins in the folds of the budding new member of the natural world.

A quick glance at my watch prompted me to move on. Time was being wasted. I still had a mountain to climb, a goal to achieve, a summit to reach, and a commanding view to capture from the top. I had to follow through with the plan.

I paused to let my mind wander and my thoughts chose their own, unscheduled detour.

I lay down on my belly in the moist soil, freshly thawed from a piercing Vermont winter. I looked at the fern and tried to imagine what it was like to break through the soil as a seemingly insignificant being in a grandiose world. The fiddlehead was tense, coiled tight into a circle like a steel spring. I stared at that first fiddlehead for fifteen minutes. Closing one eye, I looked at it through the other. I circled it and looked at it from above, trying to imagine what each part of it strove to accomplish in the first days of spring. I rolled onto my side, totally focusing on that one fern, that one sign of life, that one tiny being.

All around me, other fiddleheads in different stages of life tested the air. Some were bigger in the world, further ahead in the race for fernhood. Some had barely begun their journeys. And I knew some still struggled under the surface, like earthworms writhing their way out of the ground. I took off my watch and enjoyed the abandon of time passing unnoticed.

Ferns are a silent, ephemeral presence in the forest, but their ranks are strong. They hold a place in the ecosystems of nearly all woodlands, with between ten thousand and twelve thousand species of ferns in the wild. I know them to grow in Vermont and California, Montana and Arizona. They grace the entrances to Caesar's Palace and the White House.

Once fully grown, the existence of ferns is almost entirely passive. Their movement depends on out-side influence. They mate on air, with the breeze spreading their charm to the lucky downwind. They are delicate beacons that signal the return of life to an otherwise desolate spring forest. They are a wel-come sign and a comfort.

Ralph Waldo Emerson said that a weed is a flower whose virtues are not yet recognized. Trendy flower shops sell orchids, not ferns. Lovers give roses, not fiddleheads. Few bother to understand the rare beauty of a fern. In the office—surrounded by cellu-lar phones, faxes, pagers, and laptop computers— we move fast but often achieve little. When I play my guitar I'm reminded that it is not the notes, it's the pauses between the notes, that makes the music beautiful. Ferns and wilderness allow us a pause. A pause allows us the chance to study ferns, or appre-ciate the taste of a peach, the beauty of a smile, the laugh of a child, the smell of the morning.

Six hours after I stopped to consider the fiddle-head, I had not reached the top of the mountain. I was only inches away from my shelter, crawling through the loamy soil and watching as fiddleheads became ferns. I felt the freedom of timeless solitude.

It was not necessary to strive for a commanding view of the forest; I was better off within it, close to its denizens and their roots. I was in the neighborhood of nature.

The wild provides a laboratory and playground with questions unanswerable. It is a place where imagination conquers time, a place where the green of a fern holds more value than the green of money, a place where the arms of the creator cradle the senses, a place where ferns rule nothing yet mean everything.

Within us we have the capacity to save wilderness or destroy it. It's time to recognize our responsibility, to recognize the importance of a fern.

The Grand Canyon

While you might be able to overlook a tiny fern, it is impossible to look across the Grand Canyon without being awestruck. The mouth of a giant, the mouth of the Earth, it gapes, awaiting a sip of water from the sky. The Colorado River, its throat, runs deep in its gullet, always expanding the breadth of its smile. The walls of this mouth are spotted with craggy teeth at some points and smooth with the silky skin of slick rock at others. Its lips open wide into the face of the Earth.

Picture twenty Secret Service agents standing on the lips of the canyon, leaning tentatively over the edge and peering down into the great mouth. Their heads sway from side to side as they nervously scan the interior of the canyon for any storm trooper or terrorist bold enough to scale it. They make a funny picture.

This was the scene when I met President Clinton at the Grand Canyon. A few weeks before, the Sierra Club heard a rumor that Clinton might use the Antiquities Act to protect the Kaiparowitz Plateau, the heart of the Utah wilderness.

Teddy Roosevelt had used the power of the Antiquities Act more than half a century earlier to protect the Grand Canyon. Under its provisions, the president has the power to protect scientifically or historically important landmarks of our natural heritage from imminent danger of destruction. The Kaiparowitz Plateau qualified for its red slot canyons, historical pictographs and petroglyphs, and ancient fire rings. Native Americans used the plateau as a place to light fires before the hunt. These fire rings guided them to the hunting grounds that would sustain them. The ritualized fire rings they created became sacred places.

The imminent danger was the Andalux Company, a Dutch coal mining conglomerate that proposed to mine for low-grade coal in the plateau. Rights granted by the Mineral Lease Act gave the Andalux Company the right to excavate coal trenches. Since the coal is of such low quality, it was slated to be sent to Singapore, where lower pollution standards allow the burning of dirtier coal. For bargain basement price, a foreign company planned to gouge out one of America's natural wonders, take coal from the wreckage, ship the coal overseas, and reap all of the profit. Something fishy was going on.

The 1872 Mining Law is one of many pieces of legislation, like the Homestead Act, which helped to push people west as America grew. It is the mother of all corporate subsidies. It's the thorn in the Sierra Club's paw. The Sierra Club has tried to repeal the 1872 Mining Law since the Club was founded in 1892. The law allows companies to go into otherwise protected land and use it with few housecleaning requirements. Mining pits need not be refilled. Cyanide, often used in mining, is not monitored for leaching and the damage that it does. According to the U.S. Bureau of Mines, twelve thousand miles of rivers nationwide suffer from pollution caused by mining. More than 550,000 mines lie abandoned, with an estimated cleanup cost of $32–72 billion.

President Clinton inscribed his name in conservation history by protecting the Kaiparowitz Plateau. He trumped the Mining Lease Act using the Antiquities Act. The Sierra Club got wind of the decision at the last minute. It was September 1996, and Clinton's reputation with the Club had been jeopardized when he signed the Salvage Logging Rider. Clinton was running for reelection, and the Sierra Club was deciding whether or not to endorse him for a second time.

The cold, blustery day began with a publicity announcement at the edge of the Grand Canyon. Clinton announced he was signing the executive order that protected the Kaiparowitz Plateau. Trees were strapped into place behind the podium, standing tall to improve the photo op. The president and the vice president wore their standard environmentalist costumes. I remembered the creased flannel shirts and spotless Eddie Bauer lite-hikers they both wore from their Earth Day appearances the previous year. As I walked onstage to introduce the president, the vice president slapped me on the back and said, "We did it!" He smiled, slapped me on the back again, and just repeated, "We did it!"

I stood behind Big Blue, one of the president's three traveling podiums and, in the tradition of John Muir, I celebrated our efforts to preserve the wilderness and make the canyons, the slick rock, and the plateaus of the Escalante glad.

I started my term as Sierra Club president expecting to give eulogies for wilderness we had lost at the hands of a rogue Congress. When President Teddy Roosevelt declared the Grand Canyon a national monument, he said that we should leave it alone. Humanity could not improve upon it. We cannot improve the canyons of the Escalante; they can only improve us. Edward Abbey and Wallace Stegner taught many of us why the desert is different from kitty litter. As we banked this new national monument, a down payment on the 5.7 million acres of

Utah wilderness that we need to protect, we did so because of them.

When I was just a kid, I hiked through the Kaiparowitz Plateau and stared up with a sense of wonder. Now I know that my kids and their kids will enjoy the same sense of wonder. We were building a natural bridge to the future. How fitting it was to have a man from a place called Hope to gather us together to preserve the canyons of the Grand Staircase Escalante National Monument—for our families and for our future.

"Fellow desert rats, our landscape will survive," I concluded my speech. "As Ed Abbey used to say, 'Long live the weeds and the wilderness yet.'"

The president signed the bill with the usual pomp and circumstance. He used a different pen for each letter of his name and handed the pens out to onlookers after each left its few drops of ink on the momentous sheet of paper. After the signing we sat down to lunch.

At lunch, I sat with the president and Vice President Gore along with Roy Romer, the governor of Colorado; Robert Redford; Terry Tempest Williams, the poet of the desert; and Mike Matz, from the Southern Utah Wilderness Alliance. The president's beaming grin revealed his pride at having done something historic. I gave him a few moments before starting a conversation about the Salvage Logging Rider. "So what are you going to do about the Salvage Logging Rider?" A strained look washed over the face of the vice president. He put his head in his hands as if to say, "Oh God, why are you drilling him for more now?"

Robert Redford, seizing a gap in the conversation as the president contemplated an answer, rolled out a story about fighting fires on his property in Utah with the skill of a storyteller. Fire ripped through forest on his ranch and all hands hurled and chopped to put it out.

"You know, you're lucky," I offered. "If you lived on national forest land, someone could now come in and start logging." His property bordered national forest land that had also caught on fire. "Say good-bye to Sundance."

The president changed the subject and moved into some stories of the old days in Arkansas and grumblings about the antics of Newt Gingrich. After lunch we took a walk to the rim of the canyon, enjoying the beautiful fall day. I looked out at the awesome sight before me. Clinton kept his back to the vast open space of the canyon and continued on about Newt. As much as I tried to zone him out and focus on the spectacular nature of the place, I could not.

Finally, I interrupted him. "Mr. President," I said, my knees quaking. "I don't want to offend you, but we're standing on the rim of the Grand Canyon, a place that makes millions of people dream. And you're talking about Newt Gingrich. Shouldn't you be looking at that?" I pointed across the expanse of the canyon.

For a brief second he looked as if he wanted to

toss me, an assertive twenty-three-year-old kid, off the edge of the canyon. But he stopped talking, turned around, and looked.

He looked for only eight seconds. I timed it. Then he launched right back into Newt Gingrich. At least he looked, and he had come to protect the Kaiparowitz Plateau. I decided then that it doesn't matter why someone cares about the environment, only that they do. The president may have been thinking about Bosnia, welfare, or Chelsea. His mind may have been elsewhere, but he saved a large part of Utah's wilderness on that day.

I'm glad I don't have his job.

The Yosemite Floods: The Wrath of a River

"Dead, dead, it's just dead." Bud shakes his head back and forth from behind the counter as he surveys the empty blue pleather booths of his restaurant. He brushes a fly from the charity mints on the Formica and continues to clean aged grease spots from a grill that used to be too hot and too busy for such work.

The silence reminds me of Los Angeles after the Northridge quake. I remember leaving my ramshackle, gravity-tested house and walking out to Ventura Boulevard—seeing no one, hearing nothing except the distant wail of a car alarm set off by an aftershock, feeling a sense of desolation only possible in the wake of nature's unmistakable assertions of will.

It wasn't an earthquake that rolled through Oakhurst. Yosemite National Park was closed. The local weekly paper, the *Sierra Sentinel,* screamed a gigantic headline through the bars of newspaper stands: YOSEMITE REMAINS CLOSED! GOVERNMENT ESTI-MATES $175 MILLION IN DAMAGES. The headline was larger than the *Wall Street Journal*'s announcement on Black Thursday of the seven-hundred-point stock-market drop.

I was in Oakhurst preparing to enter Yosemite Valley, preparing to witness the adjustments of nature to the architecture of man. Nature closed two of Yosemite's three entrances, so I wound through back roads to reach the entrance near Fish Camp.

I had left earlier that day with a friend, Tom Elliott, in my zippy little global warming machine. Before I left, David Brower had made sure to mention that he had the second fastest driving time to Yosemite. Galen Rowell, a legendary photographer of Yosemite, held the record with a jaunt from San Francisco to Yosemite in just under two and a half hours. David served these two record-breaking trips up for me, almost in challenge. When I left San Francisco, I put the top down on my car and enter-tained thoughts of competing in the race. As we neared the park and were forced down increasingly convoluted routes to find access, I realized this trip was not about breaking speed records.

We were headed to see the results of the New Year's floods, the grandest and most damaging floods in Yosemite's recent history. Reports and rumors of the destruction in the valley trickled into the Sierra Club from day one. Yosemite is the spiritual homeland of the Sierra Club, the place we go to refresh ourselves when we're weary of political battles and rhetoric.

As president of the Sierra Club, I feel a historical link to Yosemite. The Club's founder and first president, John Muir, was the first white man to climb Yosemite's peaks, the first to explain that glaciers had formed the rough-hewn valley, and the first to demand that Yosemite be protected forever. He began the American legacy of wilderness conservation when he invited Teddy Roosevelt to Yosemite and convinced him to support its designation as a national park.

Past Sierra Club presidents like Phil Berry spent months of their lives in the Sierra. Their first experiences in Yosemite were as children. Initiation to the wonders of the area came through participation in the fabled High Trips. Concerned that the environmental impact of the High Trips was harming the future of the mountains, the Club discontinued them in the 1960s. Before the trips were stopped, hundreds of people trekked into the wilderness together to enjoy its challenges, character, and calm for weeks at a time.

I visited Yosemite myself a few times as a child. During my short visits, I experienced what I would later identify as a feeling of coming home—to the towering shelter of trees, the cool bath of rippling waters, the mysterious security of natural staircases leading to the dormers of the world.

But I never had an extended stay in the park. And now, as Sierra Club president, I scramble to find the time. I'm there for ceremony, assessment, and planning, not hiking, canoeing, and climbing. There is nothing more frustrating about my job.

The first report I heard of the floods came from Katie Hamilton, the fifteen-year-old daughter of Bruce Hamilton, the Sierra Club's wise and kind conservation director. Returning from a day hike on December 31, she encountered a ranger who delivered the news that her campsite was deluged, all of her gear lost. She spent the night in a paper sleeping bag with members of the Yosemite Fire Department, shivering in the fire station. The hundred-year flood had begun.

Yosemite has always been a place of great spiritual renewal. Its soaring walls and serene meadows shaped the worldview and appeased the demons of many Sierra Club leaders, Brower and myself included. In the last twenty years, we've marginalized its brilliance. The roads and hot dog stands, traffic jams and tent cabins, streams of cars—degradation erodes beauty as the valley is slowly destroyed and the nature of this magical landscape is suppressed. Humanity mimics nature as it gouges away with relentless motion like a glacier creeping through time. Unfortunately, the trail left as humanity recedes is not one of rivers tumbling over cliffs of piled stone but one of smog pouring over piles of garbage. This was the Yosemite Valley I had last seen in May with Brower.

The Yosemite Valley I visited after the floods showed a different face altogether. The place was reborn. You could see it, smell it, taste it in the breeze. Both the raw power of natural change and the serenity of its aftermath were palpable; Yosemite Valley felt real again. The floods had made it so.

Flooding is a natural act of changing a river system's status quo. Flooding provides catharsis for the mountains, therapy for the landscape. Here in Yosemite, the Merced rose up from its banks, leaving its usual bends and twists. The river joined with its network of streams as tributaries and rivulets turned into raging currents; a new temporary system was formed. Nature purged itself.

In the far eastern curve of the valley, where the Merced flows down a steep gradient and through the entrance of the wider floodplain to the west, the river rose above a thirty-foot-high bridge. Sixty-pound pieces of deadwood rammed the girders of the bridge with a force strong enough to twist steel. Bear boxes and bathroom facilities barely withstood the flow, slowing it just enough to catch the murky sand and silt the river gathered along its path from its distant origins. The river clawed at parking spaces in a nearby campground, churning up chunks of black pavement, folding it back on itself, and eventually tearing it into the current until the dirt underneath was once again fully exposed. The Merced roared up to a massive rockslide from the previous summer and changed its course, leaving its own trail of destruction in its wake. The river was not to be outdone.

Jerry Mitchell, the director of cultural resources in the valley, has worked in Yosemite for a little more than a year. Restoration is his specialty. His placement at Yosemite came in recognition of a job well done overseeing the controlled releases from Glen Canyon Dam near the Grand Canyon. After years of choking the Colorado River with this ill-conceived dam, the park service began to see the reasons that the river originally ran free. Without the steady flow of the Colorado, the Grand Canyon decayed. The park service picked Jerry to ensure that the Grand Canyon received enough of its life force to survive. A park ranger introduced him as "the man who gave the Grand Canyon its spirit back."

"How's it going?" I asked him as we shook hands on the floor of Yosemite Valley. The staff had spent the last ten days working around the clock to restore running water to the facilities. He looked like a doctor who had just come out of a long, successful surgical procedure.

Through a satisfied and tired face he smiled, "The valley is breathing again."

Later in the afternoon, after an extensive tour of the damage to bridges, buildings, tent cabins, and roads, Jerry Mitchell and I stood on one of the few remaining footbridges that tentatively spanned the Merced.

"This," he said, "was all a big lake." During the floods, the river backed up to gather its strength on its way out of the valley, forming a slow-moving body of water half a mile wide. As we approached the bridge, we saw a heavy vehicle parked in the bushes to the left. The raging current had picked it up to the east along with a cargo of pavement, tent cabins, picnic tables, and mattresses and had dropped them all here.

In my six hours there, I saw fewer than ten moving vehicles. The human infrastructure, flecks in Yosemite's geologic time, took a beating. A month after the floods, toilets had just begun to flush. "Housekeeping," a campsite normally speckled with two hundred flimsy, three-walled structures, was 80 percent destroyed. When filled in the summer, it had looked like a refugee camp. I would not regret this change.

Only the rocks remained the same. El Capitan and Half-Dome endured in all of their majestic splendor just as they had through the days of smog and hot dog stands that have recently characterized the valley. Though much was displaced, the only elements of the valley missing were the human elements. We were missing, but Yosemite was not missing us. I wondered aloud if we would be the only ones to notice how different the valley felt.

"Wildlife," Jerry responded to my question. "Wildlife will certainly benefit in the long run. A few bears may have been flushed from their dens, a few deer stranded for a few days on sandbars. But the river was set free and it removed the agitators of the valley's nonhuman inhabitants."

The whole of nature had been jumbled, mixed up, shaken and stirred. Change occurred. Without our overbearing presence, while the park was closed, the creatures of Yosemite had a free-for-all, a no-holds-barred enjoyment of home. While we stood admiring the view, Yosemite was hard at work putting itself back together, not as it had been, but as it should be. Yosemite was wild again.

I expected to see the uprooted trees, the hanging and mangled debris, the fresh river channels carved by the waters of the flood. I expected to see the flood falls pouring from the granite faces of the valley. I expected to see the ephemeral trappings of humans tossed about like so many Tinker Toys. I did not expect to see the costs the river exacted from humans outside its path. By the time I arrived at Yosemite, the park had already laid off one thousand employees.

Each year 4.1 million visitors make the trek to Yosemite, contributing $2.1 billion to the California economy. Much of this money supports the communities in and around the valley. With the park closed for the second time in a little more than a year, the communities in the surrounding seventy-five-mile radius felt the pain of suffocation.

When I'm on Capitol Hill, a place I try to avoid, I'm confronted by members of Congress who see national parks as fringe benefits for the left fringe. They believe that environmentalists are selfish yuppies who wish to halt further development in the wilderness where they already have a cabin or retreat. These are the same members of Congress who lent the Chrysler Corporation one billion dollars to save ten thousand jobs. With Yosemite closed, thirty thousand jobs in outlying communities were in jeopardy. They continue to be so as long as Congress believes that national parks are fluff. National parks are not economic wastelands, as many of our "friendly" western senators believe. They are not only an integral part of nature; they are an integral part of the economy.

Fred, an attendant I met at the Sierra Chevron in Mariposa, said it best: "No Yosemite, no gas." Once Fred's hours were reduced he was forced to apply for welfare just to subsist. He and many others feel the sting of nature's rejuvenation. He and many others rely on our use of Yosemite to survive. The question people like Fred pose to us now is this—how can we reinstitute our use of the park for the sake of visitors and surrounding communities while avoiding our mistakes of the past?

The woman the Park Service put in charge of answering the question is park superintendent B. J. Griffin, a cross between Texas governor Ann Richards and Flo, the sassy waitress from the seventies sitcom *Alice.* B.J. is always ready to scream, "Kiss my grits." She is simultaneously tough and caring, and her actions reflect deep thought. She became superintendent of Yosemite in 1995 after the tenure of Mike Finley.

When B.J. was appointed, she announced her plan to work as a partner with the surrounding gateway communities and the concessionaires in the valley. The word *partner* had meant *patsy* to superintendents like Finley. Yosemite's fans became nervous about the future character of the place they hold dear.

In B.J.'s mind, *partner* meant preserving the park without crippling the people who depend on it for their livelihoods. She had no intention of further commercializing the park. When she took control, she quickly supported the Yosemite General Management Plan (GMP).

The GMP arose out of a partnership between the public and the park. Serving as a guiding document to ensure Yosemite's survival into the future, the plan ordered the removal of campsites that sat along the river, of ramshackle garages that were eyesores throughout the park, of a gas station, and of hot dog stands. People asked why we should remove these modern conveniences. I question how we could have built them in the first place. If you want stuffed animals, go to an amusement park.

When development began in Yosemite, the average visitor stayed for an extended period. People normally drove into the park in their cars, parked them, and then hiked out of the valley to explore the wilderness. Today, the average visitor spends 4.7 hours in Yosemite—curiously, about the same time as the average stay in Disneyland.

One summer, I sat in gridlock traffic in the shadow of Half-Dome with David Brower. We sat in silence and I could see his fists clench as the Chevy Impala in front of us spewed enough exhaust to make our eyes tear. In Disneyland, people expect to wait in line at the Matterhorn. In Yosemite, people should expect to wait only for the slow and steady movements of nature. They should expect to hear more birds than beepers.

I'm not suggesting that we use the opportunity for change afforded us by nature to prevent people from visiting Yosemite. I am suggesting that we avoid loving Yosemite to death.

John Muir secluded himself in Yosemite before the turn of the century to experience wonders of natural progression. If he were alive, Muir might laugh if he heard that $175 million worth of damage occurred in the valley. "Damage?" he'd chortle. "Nature cannot damage itself, it can only be reborn."

We have a chance to give rise to a new and improved Yosemite, a Yosemite that sustains the people who depend on it without losing the qualities that make it spectacular. People depend on Yosemite for their livelihoods, and Yosemite depends on people to treat it responsibly as they explore its wonders. The floods are like a new ride at Disneyland, and America should rush to enjoy the altered park.

But Yosemite can't afford to provide the ride if each person drives his own car for a twelve-minute tour.

Okefenokee: An M&M Without the M

Whether our motivations are political, recreational, financial, or just to save nature for its own sake, we have a choice to make. Will we stand up for the natural world that sustains or will we allow it to be sacrificed so that we may enjoy conveniences like pristine white *M*s on our M&M's?

Let me make myself perfectly clear. I'm not anti-M&M. I like them just as much as the next guy. What you may not know is that M&M's are coming after the Okefenokee, the world's most famous swamp.

The Okefenokee covers 438,000 acres in southeast Georgia and northern Florida. Of the total, 396,000 acres are designated as part of the national wildlife refuge. Its towering forests of pine and cypress trees, sprawling lily-padded swamps, wetlands covered in wild grasses, and endangered species like the wood stork and the red cockaded woodpecker form a tranquil beauty. It is a place of recreation for more than 400,000 people each year. It is an ancient retreat for the humans and other animals who rely on it for a healthy existence. Whether the Okefenokee provides a purification system for drinking water or a safe place to nest, it offers its services in return for our agreement not to upset its balance.

The Okefenokee formed when the Atlantic Ocean reached its arms inland and blanketed much of what is now the east coast. As it receded, the Atlantic left Trail Ridge, a thirty-foot sand dune that forms the eastern boundary of the Okefenokee and holds the water that creates the wetlands and swamps of the area.

Thousands of years later, the Du Pont Corporation acquired mineral rights to thirty-eight thousand acres of Trail Ridge. Today, they propose to mine the area for titanium dioxide.

Because I always drifted off in chemistry class, I had to ask a chemist to describe titanium dioxide to me. Titanium dioxide is a whitener that's used to brighten our toothpaste and our teeth, to make Oreo cookie filling, and to make paint for the *M*s on M&M candies.

Although Du Pont has yet to garner state and federal permits to mine the area, they're proposing to mine a thirty-mile by three-mile stretch along the edge of the Okefenokee National Wildlife Refuge. In order to extract the titanium dioxide, they need to sift the sandy soil of Trail Ridge. First, large sections of pine forest must be clear-cut from the ridge. Then, Du Pont would strip a foot-thick layer of topsoil off the ridge. Finally, massive dredgers would be floated over mining pits to suck up Okefenokee's booty.

As soon as they heard of plans for mining in the Okefenokee, local activists sprang to arms. Josh Marks, a Georgia Sierra Club staffer, released the

Club's statement. "Our official position is this: We
still want DuPont to abandon the project now.
. . . This is a bad idea whose time should never
come." Josh gathered pictures from other titanium
mines as evidence that a mine in Okefenokee might
turn the landscape into a moonscape, spilling orange-
brown waste water into the swamp, choking plant life
and fish.

Shareholders of Du Pont stock mounted their own
protest. Forty people marched outside the Du Pont
annual meeting while shareholders inside spoke to
the executives about their concerns. The message
they sent: Choose the wild Okefenokee over mass-pro-
duced M&M's with perfectly matching white mono-
grams.

Local mobilization garnered statewide recognition,
bringing the government forward in strong opposition to Du Pont's mining plan. Bruce Babbitt, secretary of the interior, voiced opposition. He caught the
public and Du Pont off guard by attempting to halt a
development project at such an early stage. In an
announcement of his decision to oppose mining in the
Okefenokee, Babbitt encouraged Du Pont to "do the
people of Georgia and the people of the United States
who care about God's creation a favor by simply
withdrawing the proposal once and for all."

In the past, Du Pont hasn't always chosen well.
The corporation was fined in 1991 for dumping cor-
rosive acids and toxic solvents in Deepwater, New
Jersey. As recently as 1989, Du Pont was listed on

the Council on Economic Priorities' short list of worst environmental offenders for being the number one toxic polluter.

In the face of pressures from all sides, Du Pont decided to halt plans for the mining project until a "respected third party" could make sure a review of the project would be "fair, inclusive and truly collaborative." Josh Marks responded, "We don't see what any third-party collaboration is going to tell any of us that we don't already know, which is that this is the worst place they could ever put a mine."

Meanwhile, activists continue to circulate slide shows, write letters to potential political allies, and speak directly with Du Pont to find an alternative to destruction. They spread the message that the choice is Okefenokee or Oreos—how sweet the reward for choosing a national treasure over a creme-filled cookie.

The Tortoise King

In order to halt the development of Trail Ridge, a concerted effort was needed. Politicians, both national and local, citizens from across the country, and a large corporation had to be shown the importance of the Okefenokee. Often, the many forces needed to protect our country's wild places are organized by a key team that works together over years. In the case of the California Desert Protection Act (CDPA), the team was Jim Dodson, Judy Anderson, and the Tortoise King—Elden Hughes.

Elden Hughes's rippled face beams. A beard prickles over his tanned skin like tiny white cacti on the surface of a smooth and sandy desert. Long before he could grow a beard, he explored the California desert. He grew up on a small cattle ranch at the edge of the desert. When he wasn't helping around the ranch, he played in his neighborhood playground. Seven-hundred-foot sand dunes were his rolling jungle gyms. Pictographs and petroglyphs from ancient Native American graffiti artists were Elden's fairy tales.

"I'm sure I was hiking as soon as I was out of diapers," he insists with his bashful chuckle.

After attending Whittier College, Elden managed two small family businesses, his father's wholesale plumbing company and a plastics company. He then started a family. Soon, he returned to school to study data processing and worked solving computer problems, managing databases, and servicing payrolls. When he tired of that he worked for the protection of wild rivers—the Tuolumne, Kings, Merced, and Kern—and testified before Congress on protecting caves. He saw that the California desert needed help. His work became protecting the desert.

Eyes twinkling, Elden continues, "At any given moment in time, you can have only one open-ended task. All others have to be finite. If you have two open-ended tasks, you go crazy. If saving the desert

is the open-ended task, you stop worrying about
trees that are over thirty-five feet tall. That doesn't
mean you don't care about them, but at present,
they're not your problem. I don't know that I love the
desert more than the mountains or the ocean; it just
needed friends more."

In his campaign to pass the California Desert
Protection Act, Elden adhered to a few simple rules
that could be applied to any campaign.

#1. Almost anything takes longer than one would believe.

Patience is a trait for which Elden gained respect
while working to save the desert. When he began
working on the CDPA he believed it would be a short
battle, no more than two years. At the time, he was
still doing data processing and saw this as an open-
ended side project. He had no idea that it would take
eight years from its introduction in 1986 to pass the
CDPA.

The fight to save the California desert began long
before the CDPA was conceived. In 1933, when Elden
was only two years old, the government declared a
national monument at Death Valley. When Elden was
five, Joshua Tree received the same honor. While
mining and grazing could still occur around the monuments, the public and political consciousness of the
fragile nature of the desert budded.

Later, in 1946, the Federal Bureau of Land Management (BLM) was created from two other agencies and took charge of approximately 13.5 million acres in the area. In 1980, the BLM surveyed its land and found 5.7 million acres that qualified as irreplaceable wilderness. Ceding to mining and grazing interests, they proposed protecting only 2.1 million of the 5.7 million identified acres.

Judy Anderson and Jim Dodson, two dedicated desert rats, fumed. They quietly began mapping the desert to prepare for a war to protect it all. In 1986, the CDPA was introduced in Congress. Elden Hughes testified at the first hearing. The act proposed to establish the 1.5-million-acre Mojave National Park, to expand the boundaries of Death Valley and Joshua Tree National Monuments and designate them as national parks, to protect 4.4 million acres of wildlands as wilderness areas, and to acquire 490 acres of historically important land in the Indian Canyons near Palm Springs. In all, the act proposed to protect almost 10 million acres of desert, including 116 wilderness areas. The goal was keeping the most pristine portions of the California desert protected as wilderness and national parks.

Elden couldn't know how very open-ended his task would become.

#2. *In the land of the klutz, the merely inept lead.*

Elden's involvement in activism grew through his involvement with the Orange County Sierra Singles program in California. He didn't come to the Sierra Club with fire in his belly, geared up for battles with members of Congress. He was recently divorced and thought the Sierra Club would be a nice way to meet a new group of friends. It worked. He met Patty, a nurse, who would later become his wife.

Elden quickly became an outings leader. By leading hikes he met new people and showed them his favorite places in the desert.

His skill was quickly noticed. He was asked to chair the Orange County Sierra Singles. The Sierra Singles put on a dinner once a month as a way for its members to meet and enjoy each other's company. It was purely social business. Elden had other ideas.

"So you have three hundred people sitting at a table," he recalls. "You show a dozen slides and discuss the issue briefly. Have a dozen people ready with pencils, paper, a statement of the issues, and ten sentences they can mix and match." Elden pauses. His goal was to make the dating game a political affair.

Elden involved people who had come only to meet new friends. He blocked the rest room doors until the letters were written and people were ready to act. Some people had never even written a letter before. He says, "You could tell, just by placement of return address and such. Never written a letter! And here he's going to get a letter back from his senator." The early letter writing/dating games helped gain wild and scenic river status for the Tuolumne River.

Elden motivated enough people to save millions of

acres of desert. In his own words, "How do you get them moving? Let them taste success. And praise the hell out of it. I don't care if [you saved] one tree or one piece of paper."

In the land of the social singles, the merely concerned led.

#3. Leaders don't carry ropes.

When you're leading a difficult ascent up a mountain face, others need to carry your ropes for you. Jim, Judy, and Elden were able to push forward the CDPA because of their ability to recognize that others could perform some tasks better. As leaders of the campaign, they couldn't always carry the ropes. They were addressing emergencies, directing energies, and gathering strength for the final push.

Judy Anderson's early maps of the deserts would pay off later. Judy wasn't a cartographer, she simply traced the maps that the BLM had made of the wilderness areas in the California desert. At the time, they did not know that the BLM would destroy all of their original maps. Judy unknowingly captured the most accurate record of the desert.

Elden chuckles at their good fortune. "If you want to get into an argument over boundaries, maaan, you could shoot people dead when you say, 'You threw your maps away.'"

Elden took the maps to Representative Leon Panetta's (D-California) office. Panetta, later the chief of staff for President Clinton, felt the CDPA was too big. After all, it covered nearly 10 million acres of land that could be used for mining and off-road vehicle trails. Panetta's legislative aide gave Elden the news. Panetta was an important vote, too important to lose. Elden spread out a composite of the maps on which he had drawn all 116 wilderness areas included in the bill. He told the aide to point to any three wilderness areas. "I can tell you," he explained, "what resources exist, what resources are threatened, and which ones need protection, in every damn one. So how can it be too big?" The week after Elden visited the environmental aide, Panetta became a cosponsor of the bill.

He was bluffing to some extent. He humbly says, "I got to know it, the whole desert, better than anyone. Now that doesn't mean I knew each part better, but I knew something about all of it." And on any individual piece of the desert, I knew who was the expert.

Without the maps, Elden would have failed. Leaders don't climb mountains on their own. Sometimes you need other folks to help carry ropes.

#4. Let the star be the star.

At first, Elden went to Washington alone, trying to make appointments, wearing his one and only lobbying suit, and playing by all of the rules. After some frustrating trips he figured that senators were sick of hearing impassioned pleas for action. He had to distinguish himself somehow.

Enter Scotty, the star. Scotty is a desert tortoise—don't call him a turtle or you'll face Patty's wrath. The desert provides the fragile, quiet, and peaceful habitat that the endangered tortoises need to survive. If the CDPA were to fail, Scotty's home would be destroyed.

Patty casually carried Scotty, a six-month-old hatchling from adult tortoises she had adopted, around the metal detectors of the Capitol building. After several trips, the security guards began greeting Patty, Elden, and Scotty with, "Hey, the tortoise is back!" as they shuttled the unlikely lobbyist through.

When the group visited Senator Kent Conrad's (D-North Dakota) office, Elden and company quietly piled into the outer office and shut the door behind them. Before anyone could say anything, they had plopped Scotty down on the floor. Scotty began exploring. Soon, the entire office was on the floor, rolling around to get a good look and cooing at the slow-moving emissary. From interns to senior aides, one by one they were taken with the harmless and helpless charm of the tortoise. Each wanted to know how to protect him.

When Senator Conrad came out of the inner office, he scanned the room at eye level, seeing no one and knowing something was amiss. Looking down, he found fourteen of his staff crawling around a tortoise and showering it with compassion. His look went from total puzzlement to absolute resignation in a matter of seconds. He ceded, like a father sensing the attachment of his son to a scared stray puppy, "Whatever it is, explain it to me later, it's okay." He became a cosponsor of the CDPA.

Elden brought Scotty to Washington to bring his point home, with the star of the desert shining in the halls of Congress.

#5. *Don't spit in the eye of a public official or the media, no matter how small the eye.*

If Scotty strutted the official halls of Washington, it was usually the last and culminating event of a longer ambassadorial mission. He would take a road trip from west to east coast with Elden and Patty, stopping in key states along the way. Once, in Albuquerque, New Mexico, Elden hoped to have an article published about the desert and the tortoise. True to his style, he marched into the *Albuquerque Tribune* with Scotty in hand and no appointment, hoping to get an editorial written.

After patiently waiting and trying all official avenues to the editors, Elden picked up Scotty and headed for the door. He was frustrated. While twisting his way through the maze of reporters' desks, Elden was confronted by a reporter who popped his head up asking, "What do you have there?"

Elden launched into the story of the desert. He described how purples melt into whites and oranges, covering the desert's floors and walls for the few brief days of the wildflower bloom. He explained how mountains of sand, extinct volcanoes, and the world's largest Joshua tree forest punctuate the landscape. He described how the desert's nearly one hundred threatened species, including tortoises like Scotty, depend on this harsh landscape for survival.

The man who approached him turned out to be the outdoor writer for the newspaper. Elden remembers, "He got fired up when he heard about the workings of the desert. He wrote a great article. He had photographs and everything. It even ran in the *San Francisco Chronicle,* that article."

#6. *No matter how worthy the cause or noble the objective, there will always be opposition.*

The team faced powerful opposition and knew that to win overall they would have to accept some minor losses. You need to hit a few sacrifice fly balls to drive in the winning run.

When the CDPA was drafted, the claim blocks of some mining companies lay inside of proposed wilderness areas. Mining companies had rights to plots of land, the sum of which made up claim blocks. Although the commercial mining could proceed even if a claim block lay inside a wilderness area, the companies with claims preferred to be outside the boundaries of wilderness. The mining companies called Senator Cranston (D-California), the main sponsor of the bill, to negotiate boundaries.

Cranston's office called Elden, who eventually

helped draw the boundaries that circumvented claim blocks. Elden realized that he could let the opposition feel that they had won a battle without giving up any real ground in the war. He also eliminated a small battle that Senator Cranston would have had to fight and expedited passage of the bill.

When the bill was introduced in the House in 1990, a more daunting case of triage presented itself. Supporters of the CDPA needed 500,000 acres with which they could negotiate—not acres to shift to this side or that, acres that could be lost for good if need be.

"The first 200,000 acres were easy," Elden explains of the decision. "We'd already given it up. You know, if you widen power line corridors, which run through all of this wilderness, you've given up an enormous amount of acres." Widening existing corridors along a straight line didn't constitute losing lots of wilderness. The corridors couldn't be mined and paths or roads existed under the power lines anyway.

After the craftily placed 200,000 acres, they faced singling out 300,000 additional acres for possible exclusion from the CDPA. To rank the remaining acres, questions had to be answered. What area is a bill killer if they take it away? What can be given up for something of equal value in return? What can be given up for the greater good of the rest?

The wilderness areas were labeled alphabetically in order of importance. Then the letters were changed so that the importance of each area wouldn't be apparent in looking at the letter attached to it. Elden remembers with a chuckle, "We forgot. We couldn't remember which was which so we went back to lettering them in order of importance. In the end, though, only one key area was lost: the South Algodones Dunes." Although this loss hurt Elden like the sting of an amputation, he lost only one piece of the California desert in the face of determined opposition.

The California Desert Protection Act was signed by President Clinton in 1994, eight years after it was introduced. Shortly after, he issued a press release over the Internet stating, "This achievement is a tribute to the many citizens who worked with Congressional leaders and the Administration to ensure the protection of these desert gems. This act is proof that the common good and the will of the people can prevail."

The CDPA is a tribute to the many people inspired and galvanized by the efforts of Elden Hughes, the Tortoise King. "You need to be a fighter," Elden says, "a life fighter. I care about it, and over time I became the spokesperson for it largely because I was available during the daytime."

The story of the green flash isn't the story of a superhero, although it does partially explain why Elden became a superhero to the people who care about the California desert.

"Under optimal atmospheric conditions which are only partially understood," he began, "just after the sun sets, a flash or lingering green light appears in the spot where the sun was. It is most often seen in the tropics."

He paused to sift the folds of his memory. "I first read of the green flash in college and kept searching the horizon after sunset for the next four decades." After hundreds of nights sleeping under the stars, Elden had spent a great deal of time searching.

"In October of 1991 the League of California Cities was meeting in San Francisco. The eight largest cities in California had already endorsed the California Desert Protection Act, so it was unlikely the League would oppose it. But one should never assume.

"Opponents of the CDPA in the Big Bear area brought forth a motion of opposition to the CDPA. Patty and I drove north so that I could be the spokesperson for it. Pat Davison of the off-road community would speak against.

"As we drove north on I-5 we watched the setting sun. Just as the sun disappeared, the green light appeared. I said, 'Patty, we're seeing the green flash.' She said it was only a spot from looking at the sun. I said for her to move her eyes, change her perspective, and see if the spot moved. 'Oh my,' she gushed, 'we really are seeing it!'

"It lasted a full minute."

The next day Elden recited all the names of all the cities and counties that had endorsed the CDPA. The vote was twenty-three to four to reject the motion to oppose. The following day, the full league found a face-saving way to have the motion quietly dropped.

"If I would draw a lesson"—Elden slowed to think—"it would be never to get so involved that one fails to look for the green flash, or worse, so involved that one fails to see it when it's there."

SECTION ◇6◇

Spirit in Nature

Lox Isn't the Only Reason to Save Salmon

*P*eople assume that I grew up Swiss Family Robinson style, sucking on a baby's bottle made from a coconut husk. Not exactly. My grandfather's idea of an environmental experience was killing aphids on my grandmother's roses or spending a week on a chaise lounge in Palm Springs.

My grandparents did, however, instill in me the importance of giving back to the community. They would always be collecting clothes for the temple rummage sale or supporting a civil rights cause, but protecting the environment was conspicuously missing. As I became active in the environmental movement, I saw that Jews are disproportionately underrepresented in the leadership of the movement. I noticed that the Sierra Club only printed Christmas cards (until this year). Even Earth Day often falls during Passover.

Exploring my family history, I began to discover the connections between a people's history and their view of environmentalism. Jews are a case study of how religion and environmental protection have been separated for too long.

As I became more and more interested in the environment I could not understand why my family wasn't as moved by the issue as I was. I argued endlessly with my father and questioned my grandfather. Finally, my father showed me a precious old book that contained writings from his father. It was the story of his last days in the shtetl of Felsteen, a town in the Ukraine.

Sunday, February 15, 1919:

> *The mood in the shtetl was very depressed today on account of the news that the city of Proskurov was annihilated as a Jewish city. It became a river of Jewish blood, a bridge of Jewish bodies. . . . Everyone walks around dejected in fear of what tomorrow will bring.*

Early Evening:

> *All assembled at "The House of Study" in order to pray, including those who never set foot there before. The place was very quiet. One could only hear the quivering sad voice of the cantor and the quiet mutter of the assembled. Each one prayed not with his lips, but with his heart. . . .*

Night:

> *Tonight, no one slept. Every now and then another*

curtain was lifted and quickly lowered. The night passed in terror and fright. It seems that the night is years longer than the day. With impatience everyone waited for the day to come, maybe there will be some better news.

Monday, February 16:

Today came and brought some news, not good news but some details of the slaughter of the Jews in Proskurov, which made one's hair stand on end.

Early Evening:

On the street there assembled a few circles of Jews talking about the Proskurov tragedy. All of a sudden there were shouts: Jews, save yourselves! The murderers are here! The destroyers are here! We are all lost! *There was a great tumult, great hubbub, screams and running. Children lost their parents, brothers, their sisters. They ran around in circles seeking each other. From some homes people started running toward the fields. All of a sudden it became so quiet, as if no one remained in the shtetl.*

The Petilura, Ukrainian soldiers, went through the center of town, singing that they will destroy all the communists, and disappeared.

Tuesday, February 17:

Horror also has a limit. When it comes to the last degree, a person loses completely his fear and remains stoned, indifferent to the worst of happenings.

On the call of trumpets, the Petilura garrison gathered. A complete stillness reigned over them. Many of them kept on looking on all sides, like ani-

*mals before they attack their victims. Soon came
their leader, the Ottoman, for whom they waited so
patiently. With sparkling eyes, they turned to their
leader, with enthusiasm they listened to his speech,
in which he gave them courage to do their "holy
work" without fear. . . . After the priest's prayer to
God, to help them in their worthy cause, they all
crossed themselves, and like animals attacked all
Jewish homes for murder and pillage.*

*The town of Felsteen had a Jewish population of
2,200. 1,100 were slaughtered without resistance.*

When I read this I began to understand why the
environment seemed like an unimportant cause to
my grandfather. When you fear annihilation, hiking
trails are the last thing on your mind.

Protecting the environment is more than just saving hiking trails. I stumbled on a history book from
the period. During this time in the Ukraine, corn
production began to decline. The topsoil eroded from
overfarming and the lack of biological diversity. Soon
the wheat fields became barren as well. The whole
system suffered an ecological crash. When resources
become scarce, it is the poorest members of society
who suffer first. The Jews, the day laborers in the
fields, were blamed for the farm crash. The White
Russian troops came after them looking for revenge.
In an environmental catastrophe like a regional ecological crash, no amount of civil rights protections
will save society from its worst tendencies. The
weakest members of society will suffer. In the 1980s,

more people became refugees due to environmental causes than political causes.

Because Jews today fear that the social climate could lead to this persecution, civil rights organizations are an obvious place to focus resources. The feeling is that if the civil rights laws of a country are strong, neither Jews nor anyone else needs to worry about persecution. Civil rights laws alone can not protect your rights. If the land is destroyed, laws become meaningless. Still, many Jews view the environmental movement as a peripheral movement of luxury.

It's time for Jews to step up their own personal responsibility for the global environment. I know I owe it to my grandfather.

Riding Llamas for God

Christians, like Jews, have yet to embrace the environment as a religious concern. A few months after I was elected, the United States Junior Chamber of Commerce honored me with an award for being one of "Ten Outstanding Young Americans," proving that you can fool some of the people all of the time. Ralph Reed, the thirty-four-year-old former executive director of the Christian Coalition, also received the award. I had the pleasure of meeting him at the award dinner. He arrived in a well-tailored blue suit, a white shirt, and a tie that quietly communicated he had always known he was just right. The tie didn't take any chances. The red and blue stripes hung there as if they were the only way that they could be. I was looking forward to meeting him. We both get patted on the head enough times to get whiplash.

When we were introduced, I asked Ralph in a matter-of-fact sort of way, "Did you notice that the Christian Coalition supported, almost to the letter, the exact opposite candidates than the Sierra Club in the recent election?"

"Well yes, we did," he replied.

"And did you notice the result in the cases where our candidates ran against each other? In almost every case we won." I had to rub it in a little. I admit, I was gloating.

"Yes, we did notice this," he said, beginning to scan the room for friendly faces.

I continued, "What's un-Christian about being environmental?"

Ralph Reed and I don't agree on many things. The environment is one issue on which we should agree. Scripture is strongly pro-environment, with many passages like Leviticus 25:23–24, where God says, "The land is mine and you are but aliens and my tenants. Throughout the country that you hold as a possession, you must provide for the redemption of the land."

In light of passages like this, the Christian Coalition's support of anti-environmental candidates is troubling. Conservative Christians should be outraged by the recent Republican onslaught against our fundamental environmental protections. Both the Christian Coalition and the Sierra Club support profamily politics, the Sierra Club in that it seeks to protect our environment for the health and enjoyment of families. Although we worship in different places with different words, we have an opportunity to work together to protect something greater than any one of us—God's beautiful creation. Our children deserve no less.

My meeting with Ralph Reed inspired these thoughts and, shortly afterward, I expressed my ideas in an op-ed article. Needless to say, members of the Christian Coalition weren't pleased. At least not the ones who wrote to me. This one's my favorite:

> I'm so grateful for Sierra Club National President Adam Werbach's enlightening letter. I had no idea that the Christian Coalition was causing all this pollution and killing people and animals by misinterpreting the Bible.
>
> For years I mistakenly thought the pollution was coming from all the cars, trucks, tools, appliances, machines, air conditioners, buildings, and factories owned by the 600,000 Sierra Club members and all their corporate contributors and favorite elected officials. And I thought our children and old people were being murdered by the abortionists, drunk drivers, and diseases being spread by the addicts and homosexuals in the Sierra Club. I thought the hubris that Mr. Werbach refers to was concentrated in his Sierra Club. Well, I stand corrected.

I was pleased to see that at least the writer knew
how to be sarcastic, but I was once again reminded
that environmentalists and Christians communicate
as effectively as the Three Stooges.

I once stated that we need a paradigm shift in environmental thinking. In the face of shrinking wilderness and increasing numbers of asthmatic children, it
seemed appropriate to suggest that a better example
be set for the way we treat our environment. An evan-
gelical friend of mine later told me that it is exactly
this type of statement that offends the Christians who
believe that only God can define paradigms. I never
dreamed that my line of thinking was offensive.

From its start, the environmental movement often
criticized organized religion for using Scripture to
justify the damage inflicted upon our environment.
Jews and Christians defended anthropocentric acts,
holding that dominion demanded the destruction of
nature to serve humanity.

In the 1960s, when antistructural sentiments characterized the environmental movement, the animos-
ity between organized religion and the environment
grew. Environmentalists lived up to their stereotypes
of underarm-hair-growing, beard-wearing, Abbie
Hoffman–following hippies. They were so busy reject-
ing everything that they didn't pause to think that
religion could be a useful tool to help achieve their
personal and political goals.

There's plenty of blame to go around for the lack
of cooperation in the past. It's time to move on. We

need to make the powerful connection between people of strong religious faith and those dedicated to environmental protection.

The story of Noah's Ark tells of how God cares for creation and serves as God's call to humanity to protect all of creation.

The waters of the flood were imminent. The world was to be destroyed because of the evil of humanity. But God found grace enough to save humanity through Noah, and to save the creatures of the wild through humanity.

When God commanded Noah to build the ark and to gather the animals, Noah was not only asked to save the pretty animals, or the fuzzy animals, or the animals he could breed and be sold for a profit. God said, "And of every living thing of all flesh, two of *every* sort shalt thou bring into the ark, to keep them alive with thee" (Gen 6:19). The Endangered Species Act was born, and on it was listed every living thing of the Earth, including humanity.

Peter Illyn, a Christian minister and the Northwest regional director of Green Cross, is more concerned with opening minds than casting blame. He says, "We've got a message. The message is simple. The Bible says taking care of the Earth is the right thing to do. If there's only one thing I can say, only one conversation I can have with someone, if I'm going to leave one impression, that's it."

It seems strange that a minister would offer an environmental concern as the single issue he would ask people to address. In his mind, concern for the environment is part of a greater concern connected to many other issues of caring for God's creation. "I believe in the creator," he states without doubt. "As we honor creation, we'd better honor the creator. By getting involved in the environment, the faith of the flock will be strengthened.

"We engage people in two ways," Peter explains of his environmental ministry. "First, we make the natural world precious in their hearts. We make them see it as something of beauty like an innocent child. Second, we try to connect interested pastors—to bring them together in a gentle way."

Peter takes people directly into nature to experience its power and understand it as a reflection of God. He developed this method of engaging people while in the wilderness, trekking one thousand miles along the Pacific Crest Trail with llamas. "After nine years of preaching I took a sabbatical to get lost in the wilderness. In the summer of 1989, I bought two llamas and kissed my wife and two kids good-bye."

He remembers his departure as bittersweet, more bitter than sweet at first. "The first couple of weeks in the woods I was angry and frustrated. I was working a lot of the stress of life out—ministry is full of ambiguities, it's full of camps and factions."

His trip took a turn for the better as he began to experience the restorative qualities of nature. "I took the Pacific Trail. It was my first extended trip into the wild. There are rhythms you find in the wilder-

ness and you can't find them in three-day trips. I was
feeling in the forest a deep, deep peace. Godliness
with contentedness is great gain." Peter mixes
Scripture, in this case words from Timothy, into his
explanation of his wilderness experience.

Peter gradually felt the Scriptures unfolding all
around him. "I began to realize there is a natural
goodness in the wilderness. Everything in the forest
was doing what God wanted it to. I didn't have a lot
of reading material, so I read the Bible from [an
environmental] perspective."

It was here that Peter found a clear biblical call to
environmental stewardship. In Ezekiel 34:17–18,
humanity is questioned, "As for you, my flock . . . Is
it not enough for you to feed on good pasture? Must
you also trample the rest of your pasture with your
feet? Is it not enough for you to drink clear water?
Must you also muddy the rest with your feet?" And
in Genesis 1:26, humanity is charged with stewardship: "Let us make man in our image, and let them
rule over the fish of the sea and the birds of the air,
over the livestock, over all the earth and over all the
creatures that move along the ground."

Peter cites these lines, saying, "We don't debate
dominion, we've got it. Everyone knows dominion, but
no one follows the Scripture that holds us accountable." The argument that God created humanity to
rule over the animals and subdue the Earth has been
used in the past to justify the anti-environmental attitude of Jews and Christians. A closer look at God's

edict shows that it calls for humanity to be environmentalists.

In the Judeo-Christian tradition, humanity is not
only a part of creation but also set apart from creation to manage it with the intents of God firmly in
mind. In this dual role, humanity is called to rule
over part of itself in ruling over nature. Given dominion over yourself, do you recklessly remove limbs?
Do you drink poisons carelessly?

Skeptics say that because God destroys nature
through floods, hurricanes, earthquakes, and other
natural disasters, humanity has the right to follow
this example of destruction. What this argument
ignores is that God, as caretaker of creation, also
revitalizes and renews nature. A gardener breaks
the soil with a hoe in order to let new life grow. If we
dam a river and the life it sustains, we don't do the
momentary damage that paves the way for rebirth.
We permanently change that which represents God
and that which God has placed in our care.

When Peter Illyn returned from the wilderness
with new insight into the Scriptures, he didn't go
back into the ministry. Instead he became a llama
guide, still anxious to show people the powers of
God's creation through direct experiences in nature.
He took all types of leaders, from Nike executives to
local pastors, out to the wilderness.

"I could take church members until the Pope
becomes a Mormon and that won't change anything,"
Peter notes. "I take leaders." In the course of these

trips, he says, "I began to see how people's lives can be refreshed and renewed by time in the outdoors. I began to see that the wilderness experience was being negated by the harvest mentality—the belief that the only good tree is a harvested tree. Trees are being trees, and by being trees they are honoring and worshiping God. [The feeling from being in the wilderness] is the same thing that you feel after a powerful church service, after you feel the presence of God.

"We wipe ourselves with the choir of God," Peter observes of our trees. "That's how I describe using toilet paper made from ancient trees. Headwaters Forest is the oldest living member of the choir of God, the oldest living thing to sing God's praises. Do we need to cut it down to make shingles? Is ours the only voice in the choir?"

When he was a minister, Peter took troubled kids on camping trips. Sitting in the forest offered them an opportunity to "get down to the nitty-gritty and understand." Now he has expanded those methods of connecting people with nature.

Peter engages people in nature through a program with the National Forest Service. He enlists Christians to work in partnership with wildlife biologists and park rangers to monitor great gray and spotted owl nesting sites.

As partners, volunteers drive to predetermined calling stations in the Gifford Pinchot National Forest where they hoot for owls at night. Once an owl responds, the volunteers return during the day to locate the nest. The partners then agree to return to the nesting site at least twice during the year to determine if the nests still exist and to see if fledgling owls have been born. In this way, they help wildlife biologists and park rangers who suffer from cutbacks in funding of the National Forest Service.

The mutual benefits are clear. Peter explains, "They get a pool of volunteers who understand the complexity of the job and the challenge of funding the Forest Service. They also get a constituency for environmental protection." As for the volunteers, Peter simply says, "You can't sit in an old-growth stand watching an owl without feeling a connection to the wilderness—and, as a result, a connection to God."

This cooperation between volunteers and the National Forest Service provides one isolated example of how Christians and environmentalists can benefit from cooperative efforts at preserving and restoring our environment. After all, we both preach, use bumper stickers, and always seem to be asking for money.

"The environmental movement has been championing creation for a long time. They can be the bell ringers." Peter says. "They've been accused of being alarm ringers, but even their shrieks are science based. And from the environmental movement, the faith movement can understand the call to action. I'm an old hitchhiker and my theory is get in, find common ground, and get along. You don't have to marry their daughter."

Eighty to one hundred million people in the United States, one-third of the population, describe themselves as Bible-believing Christians. Overall, four-fifths of Americans identify with the Judeo-Christian tradition. But Americans identify with other religions as well, most of which contain similar connections to the preservation and careful management of our world. The "greening" of these religions is occurring throughout the world.

In Asia, Buddhists established the Buddhist Protection of Nature project with the support of the Dalai Lama. The project promotes Buddhist teaching about the environment to raise awareness. Buddhism not only emphasizes that humanity is part of nature, giving its adherents a sense of respect for it, it also adheres to the idea of reincarnation. For a Buddhist, abusing an animal could mean beating a dead relative. Dharma, a central idea in Buddhism, focuses on the teachings of Buddha and the nature of all things, stressing the interconnected nature of everything in the universe.

The Free Tibet movement, one of the most important and visionary social movements today, incorporates environmental activism throughout their agenda.

In Syria, a monastery is being rebuilt to provide a meeting place for Christian–Muslim dialogue and environmental studies. Adherents to many African and Native American religions view the gods as indistinguishable from nature and value their natural world as godly. A Hindu text, the *Rig Veda,* explains creation as having sprung from the parts of the Purusha, the primordial being. As such, in Hinduism all parts of the environment are held sacred.

Preserving and protecting the environment offers the opportunity for cooperative efforts between the people of the world. And a vast cooperative effort can lead to an environmental success that invigorates peoples of all faiths, investing them with a new hope for humanity and its ability for positive advancement.

Environmentalists and religious groups often have the same goals. We must reach out to each other, working together in our common mission to protect the natural world, to protect God's creation, and to protect humanity as part of the natural world and steward of God's creation.

Why We Must Drain Lake Powell

Step for a minute into Glen Canyon, the most mysterious and beautiful stretch of canyonlands along the Colorado River. It is a place that you cannot see today. Because if you could, this is what you would see.

You face a wall of rock. Gracefully and quietly, it reaches hundreds of feet into the distance above you. Your eyes begin at the slick and shining pebbles at the base. The pebbles tell of a time when another layer of porous rock, somewhere upstream, fell from its perch under the weight of absorbed water. Now these remnants of nature's sculpture, scarred and smoothed from their journey, rest peacefully in the shade of an overhang.

From the pebbles your eyes trek upward, crawling up a wall striated with tan, brown, black, beige. Yellow washes away almost to white where the sun points its most brilliant fingers. Your eyes search up the weight of the wall until they pull your head backward. Here, the rock arches up and over you, throwing itself another hundred feet straight out, perpendicular to the towering face, suspended by imaginary wires that hang taut from the crisp skies above.

You are enclosed in an alcove. The walls block out all intruders except the twisting sunlight. The ceiling feels as if it may fall at any moment. The whole of the rock is like a diver, suspended at the moment her feet are about to leave the diving board. Her back arched and ready to tumble into a series of flips, her feet never more firm than at this crucial moment of take off, she is time. Both in motion and as unmoving as the rock that forms her, time stands before you, above you, around you.

The place you admire is the Cathedral in the Desert, only one of the wonders of Glen Canyon. It is aptly named, spiritual in that it sets all of your deepest senses free while taming them with humility. It defines and defies time.

It took time to create this wonder. Two billion to half a billion years ago, the Precambrian era shaped the bedrock of this spiritual haven, hardening sedimentary mud into rock. None of this can be seen in Glen Canyon, but it is the foundation on which the entire 186 miles of canyonland is built.

Then, during the Paleozoic era, which comprised the next three hundred million years, great layers of limestone plastered the foundation. The stucco floor was laid down in a time when the edge of the Pacific Ocean lapped at the banks of what was to become Utah.

The land rose, pushing back the waters of the ocean. Mountains arched their sleepy backs in Nevada and western Utah beginning in the Jurassic period, about two hundred million years ago. And for another fifty million years winds whipped across the burgeoning plains and deposited the sand dunes that we would later name the Colorado Plateau. Shallow seas, unsure of their tenure, came and went for the next seventy million years, washing the immense dunes of sand. Swamplands and tropical forests played host to the dinosaurs, later becoming deep coal deposits.

Between the Cretaceous and Eocene periods, nature continued her task. A shifting of continental plates pushed up the Unita and Rocky Mountains while cracking and rearranging the sedimentary rocks that form the foundation of the canyon. Then, ten to fifteen million years ago, more shifting of the earth's crust shot the entire area upward and created rivers like the

Colorado. These new, strong rivers washed away deposits, cut into sandstone, and created more than two hundred side canyons and an array of splintering slot canyons that together make up Glen Canyon.

In less than seven years, humanity defamed two billion years of nature's work. On April 11, 1956, Congress authorized the drowning of Glen Canyon. On October 15 of the same year, the president of the United States pushed a button in the Oval Office, triggering the ceremonial blast that heralded the beginning of construction. Eight months later, the first bucket of concrete began choking off the Colorado River. Three and a half years later, the last bucket of concrete provided all that was needed to throttle the river and fill Glen Canyon with twenty-seven million acre-feet of water.

Back in the heart of Glen Canyon, imagine wandering through a maze of slot canyons. On either side of you slick rock walls point high, admitting only a slight wash of sunlight between their narrow gaps. Any light reaching you hundreds of feet down in the tight passageway dissipates as it filters through the confines of tunneling space. The limited light works its way into these deep crevices and catches inconsistencies in the smooth face of the rock. Suddenly, a spear of mineralized red shoots down the wall only to be absorbed by the surrounding blackness after a few feet of travel. Upon closer inspection you see brachiopods, geologic signposts of the past. Occasionally, the gap between the rocks widens and the brown spotted with streaks of yellow reveals itself again.

From above, the cry of a red-tailed hawk dips down into the rock, bouncing off slanted walls until it gently taps the skin of your eardrum. Struggling now in a slot canyon that has narrowed to only a few inches, you reach up to pull yourself over an impasse. You are crawling through the cracks of time, unsure if they lead you to the core or to the surface. The only way from here is up, yet your senses fail you. You cannot know for sure how to reach the top. The blinding light shines at the end of the tunnel, hundreds of feet above, but your human form is trapped for now with feet bound to earth.

Then you emerge into a more open space. The high slots of a narrow canyon give way to the gentle slopes of Seven-Mile Canyon. The new walls reach from top to bottom like canvas awnings. Stripes of limestone white and mineralized teal, the tracks of water seepage, alternate along all faces. A new kind of sunlight, bold and unfettered, washes over the scene. From confinement and doubt into vastness and ease, you encounter the peaks and valleys. Trials and rest await you, the two continually giving way to each other.

Again, the sense of time prevails. Time used for awkward movements through slits in the wall. Time spent in admiration. Time predicted to negotiate a new terrain. And the time of the river, to which you have returned, leading you through the wonders of Seven-Mile Canyon.

Some fought to save Glen Canyon, but in the end it was sacrificed to prevent the Echo Park and Split Mountain Dams from flooding Dinosaur National Monument in Colorado. By 1950, the obvious sites for dams in the United States had been identified and utilized. The rivers of the eastern United States already bore these harnesses of humanity, and the rivers of the west, specifically those near the coast, were yoked as they were discovered. The Army Corps of Engineers and the Bureau of Reclamation, aggressive agencies in the Department of the Interior, looked to the interior regions of the country for new projects. Hungry for fat pork-barrel projects to puff up their budgets, they sniffed out the scent of the Colorado River.

With low rainfall, harsh winds, and a short growing season, the Colorado Plateau makes a poor spot for cultivation and agriculture. With its twisting network of steep-walled canyons, visionless men trapped in the juggernaut of progress saw a perfect location for dams.

Unable to justify new projects in the area under the simple premise of promoting agriculture, the Bureau of Reclamation devised a plan for "cash register" dams—dams that would earn money by selling electricity. The profit from power would then underwrite any development and agriculture plans. Two "cash register" dams were proposed to flood Dinosaur National Monument.

One of the dams, the upper dam at Echo Park,

would provide the peaking power needed in times of surge. Pumping in surplus power, this dam could enable all of the citizens in Salt Lake City to turn on their hair dryers at once. It could also send fantastic surges of water through the canyon that destroy wildlife habitat, rip up the tamarisk and Indian rice that grows along the river, and wash out the sandy beaches on the shore.

In order to prevent some of the damage caused by the first dam, the Bureau of Reclamation proposed a second dam at Split Mountain, where the river rushes through a mountain, splitting it in two. The second dam would catch surges of water produced by the first in an attempt to limit damage downstream. It would also release water at a more even rate, providing steady "base-load" power and greater revenues.

The Sierra Club mobilized to prevent the Upper Basin Project, which included the dams at Dinosaur National Monument and the Glen Canyon Dam. Dams became a rallying symbol for a generation of activists. The Club geared up for a campaign of education with traveling slide shows, and prepared to bring the message directly to politicians.

To educate, the Club brought people to the site of Dinosaur National Monument. From ordinary citizens to journalists and politicians, people rafted through the canyons of Dinosaur. At the time, most had never seen Dinosaur and believed its importance lay merely in the fact that it offers one of the world's greatest

stores of dinosaur fossils. Few had any idea of its scenic beauty, comprising expanses of slickrock, delicate flowering cottonwoods, and maidenhead ferns. Once you saw it, you'd fight to protect it.

To bring the message directly to politicians, the · Club sent concerned citizens to speak to committees in Washington, D.C. In January 1954, David Brower went before the House Committee on Interior and Insular Affairs to plead the case for Dinosaur. He disputed the Bureau of Reclamation's water calculations, a bold move for someone without a degree in science.

The bureau claimed that the three dams, one at Echo Park, one at Split Mountain, and one at Glen Canyon, would best conserve water. Brower pointed out that one higher dam at Glen Canyon would cause less water to evaporate than the combination of the three smaller dams. He actually advocated the building of a dam. Environmentalists were shocked. The greatest conservationist of the era made the tactical choice of a lifetime.

Brower was proven right, at least about the figures; the bureau miscalculated the surface area of the three reservoirs to be created by the three dams. In the end, he was also forced to live by the deal he proposed, and Glen Canyon was lost. The dams that would flood Dinosaur National Monument were dropped from the project list of the Bureau of Reclamation. Instead, the bureau focused its efforts at plugging up Glen Canyon.

While you explore the wonders of Glen Canyon this time, your diving mask tugs at the hair on the side of your head. A small pool of water collects in the bottom of the mask as you adjust it. Your dry suit, necessitated by the cold lake water, weighs you down in the world you explore.

No sunlight reaches deep enough into the submerged canyon to illuminate the tapestry of colors before you. All is mottled grays and blacks as you weave through an underworld of dark and mysterious caverns. Much of what you came to see lies beyond your reach, hundreds of feet below the surface of the lake. The walls sag and sometimes crumble in massive sheets with just a brush from your flippered foot.

As your air bubbles escape and shoot to the surface, they are met by a slick of gasoline, mistakenly spilled in Lake Powell. Muffled sounds of engines reach down into your gloomy expanse, a chilling noise in a dark and silent world. A look above provides no blue sky to complement the browns, greens, and purples of wet slickrock. Instead, as you head to the surface, you see the white whir of a houseboat propeller and the synthetic royal blue of hullpaint.

The spirituality of Glen Canyon is saturated and stifled by humanity. No longer do colors wash into each other while the skins of rocks sweat. The light does not fight its way to the bottom of a long and vertical tunnel, touching streaks of flaming color on its way. Natural awnings don't spread themselves out along miles of canyon wall. And in no place does time stand still, as water seeps into the walls and silt accumulates from floor to ceiling.

David Brower only found opportunity to experience Glen Canyon just as the waters of the Colorado began backing up in it. He later wrote to the public, "Glen Canyon died in 1963 and I was partly responsible for its needless death. So were you. Neither you nor I, nor anyone else, knew it well enough to insist that at all costs it should endure. . . . So a steel gate dropped, choking off the flow of the canyon's carotid artery, and from that moment the canyon's life force ebbed quickly." The lesson: Never sacrifice something because it is not known to you.

Today, Glen Canyon lies submerged under water that is 560 feet high at the dam face. Glen Canyon Dam, the second largest in the Western Hemisphere, is constructed of five million yards of concrete. In some respects, it heralds the achievements of humanity.

Lake Powell, the reservoir created by Glen Canyon Dam, fills 186 miles of the Colorado River gorge. The coastline of the lake, following the submerged side canyons, is longer than the western coastline of the United States.

Although the lake provides a recreational spot for more than three and a half million people each year, it has been humanity's most successful experiment in waste. Between the annual 700,000 acre-feet of water lost to evaporation and the accumulated eighteen million acre-feet lost as water seeps into the walls of the submerged canyon, estimates put annual loss of water above one million acre-feet per year, about 8 percent of the flow of the Colorado River each year. If we recovered just one year's loss of water, we could supply the water demands of Salt Lake City for five years. For a thirsty desert Southwest with a fast-growing population, the water is too valuable to waste.

Even worse, Lake Powell is a temporary reservoir. The Glen Canyon Dam not only stops water from passing, it also catches all of the suspended silt that the water carries. In what may be as little as one hundred years, silt will cover the four outlet tubes that control excess flow. When this happens, water may rise out of control, causing massive flooding and damage downstream throughout the Grand Canyon.

As if this waste and potential destruction were not enough, the dam and lake already exact their toll downstream. The Colorado River no longer reaches the Pacific Ocean, partly due to the Glen Canyon Dam. The life along the river in the Grand Canyon continues to die off, as it no longer receives the normal flow of the river. The average temperature of the water that does reach the canyon is forty-seven degrees, too cold to support the life it should sustain. Wildlife fights an uphill battle just to survive. I don't want to be remembered as part of the generation that destroyed the Grand Canyon.

Nor do I want to suggest that there is nothing extraordinary about Glen Canyon Dam and Lake

Powell. Dangling Rope Marina, one of five gas depots on the lake, pumps almost two million gallons of gas a year, the greatest volume of any Chevron station in the world.

Glen Canyon is not yet totally lost. The Sierra Club board of directors voted recently to advocate the draining of Lake Powell. When I placed the position on the board's agenda, I didn't expect the board to have the courage to pass it. I was wrong. The board voted unanimously in favor of the position, stating loudly that it was time to right a wrong that the world was too slow to recognize.

The Glen Canyon Institute, founded by Rich Ingebretsen, an emergency-room doctor and a member of the Church of Latter-Day Saints, has helped lead the charge. It offer river trips through similar nearby canyons to show people what awaits us underneath the waters of Lake Powell. They hold educational forums to discuss the issues involved in decommissioning dams. They teach classes to share information about Glen Canyon and about the use of water from Lake Powell. They work to design, advocate, and implement policies for the restoration of rivers throughout America, with a focus on the Colorado River and its stuttering trip through Glen Canyon.

Many think that Glen Canyon can never be the same. Of course they're right. Nothing that endures exceptional hardship remains the same. The hope resides in the old saying: That which doesn't kill us makes us stronger. While the upper parts of the main channels in Glen Canyon are certainly silted in, the damage to the canyon as a whole at this point is minimal. It is difficult to predict how forty years of water will have affected the walls of the canyon deep below the surface of the lake, but much of the rock will resist its effects. Ecologists estimate that the plant and animal life at the banks of the river would be completely restored in the first year. The "bathtub" ring that will show around the lip of the canyon will take longer to fade, but should be gone in just a few decades, in time for your grandchildren to take a raft trip down Glen Canyon.

If the lost city of Atlantis were discovered, we would certainly try to uncover it and to know it despite any damage that may have been done. We already know what lies beneath Lake Powell. Its beauty alone, not to mention the three thousand Paiute archaeological sites and the canyon's latent spirituality, provide us with ample reason to uncover it.

The best reason to reclaim Glen Canyon is hope. If we can reclaim Glen Canyon from the jaws of our past folly, we can restore the rest of the destroyed environment. No problem is too large. No challenge is insurmountable. Tom Hayden once said that all he's been able to do in his life is slow the rate at which things get worse. That's not acceptable to me. This is our chance to move beyond putting Band-Aids on problems. This is our chance to make things better.

Following Brower's example, I will dust off my dry bag and lead small groups of caring people down Cataract Canyon until Glen Canyon is restored. Cataract, the canyon immediately preceding Glen Canyon, gives a partial sense of the majesty we could reclaim. Once you've seen it, you'll understand why we need to save Glen Canyon.

You're back in Glen Canyon, where Rainbow Bridge arches its natural stony back hundreds of feet into the air. Like the gateway to some mystical city, it opens wide a passage to peace.

Around another bend, Coyote Bridge frames a solitary grove of cottonwoods. They shoot upward while hanging mosses reach down to them from cliffs above. Scrub brush and ferns finish the idyllic picture that seems to contain the origins of life itself.

Progress down the length of the river reveals the flow of water in Coyote Canyon. Nature takes its varied course after a flash flood, sending white waters over the layered shelf of rock. The white hangs like a wedding dress over the dark tuxedo of the rock. A marriage of water and rock provides evidence of the timeless rituals of this place.

Water, rock, life, and time unite under the surface of Lake Powell. The elements of spirituality band together in the depths of Glen Canyon, waiting to be unleashed, awaiting their reunion with humanity. Glen Canyon holds no grudge and asks for no penance. It simply awaits the prodigal son with open arches, with regalia in all of the colors of the world, with the sounds of many tongues, and with renewal in the kiss of a river.

The Ruby Repairer

Ages ago, there was a castle called Fondue ruled by the merciless King Muenster. The king was a cantankerous, drunk, stinky old man with dark spots all over his skin. He stormed about the castle trying to escape an odor that mysteriously followed him from the top of the tower to the depth of the dungeon. When he spoke, he bellowed from the pit of his stomach. His subjects cringed in fear as they pinched their noses to block his awful stench.

After King Muenster wrested power from the mild and complacent King Gouda, he beheaded all of the castle's workers except for Pumpernickel, the senile old cleaning lady. Pumpernickel was greatly loved by the people living in Fondue because she mopped up the drips and splatters. Someone was always making a mess in Fondue. Unfortunately, the darling old Pumpernickel had a taste for cheap wine. She fell into things constantly.

When first King Muenster met her, Pumpernickel was busily mopping up a spill on the floor of the dining room and she bumbled into Muenster with her greasy napkin.

"Oooohoh," she cackled with her crusty voice. She steadied herself and placed the tines of her glasses back behind her ears.

Muenster, angered by the brush of her napkin, turned to her and grunted, "I'm keeping an eye on you." He slid away from her, looking again for the source of the odor that followed him.

When he wasn't searching for some olfactory relief, King Muenster rubbed his favorite possession, the giant royal red ruby. The ruby pulsed on the castle wall, burning bright on melty summer days and taking on a sanguine hue when the weather cooled and the people of Fondue lazed around motionless under thick coats.

One day as Pumpernickel busily wiped up the king's quarters she came upon the giant royal red ruby and began to wax it with her greasy napkin. Just then, King Muenster walked in. The putrid scent of him startled her and she swung around on wobbly feet like a newborn fawn on a patch of wet grass. As she flailed, she flung an arm and knocked the giant royal red ruby right off the wall!

King Muenster globbed down onto the floor, trying to break the ruby's fall. But he was unable. When he reached the giant royal red ruby he picked it up to examine it. In its core he found a thin, spined crack running from the very center to the bottom point. He wept rancid tears.

"Pumpernickel!" he fogged, regaining his composure and sending an anesthetic dose of breath into the air. "Give me one reason not to behead you!" King Muenster already had Pumpernickel by the wrist, ready to drag her stale old body to the depths of Fondue.

Pumpernickel crumbled to her knees to beg. She begged for forgiveness. And as she knelt, an idea arose in her mind.

"My son Jack!" came her crusty cackle. "He can repair the royal red ruby for you."

At the thought of Pumpernickel's ignorance the king spat, "No one can repair a ruby, you dolt."

Pumpernickel pleaded. "Please, sire, please allow him to try. You have nothing to lose. Give him one night. And if he doesn't please, you take my head."

King Muenster agreed. He was intrigued by the confidence of Pumpernickel in her scrawny, middle-aged son. So Pumpernickel brought the giant royal red ruby to Jack, who spent the whole night working his magic with it.

The next morning, the king walked into the outer room of his chamber and saw the giant royal red ruby once again in its fixture on the wall. He was shocked. He rejoiced, surprised that she had told the truth about her son.

King Muenster rubbed the giant royal red ruby, thrilled that it felt like new. Jack had polished it to a brilliant luster. But when the king looked closely at the ruby, he saw that the crack was still there. Fuming, he sent for Pumpernickel and her miscreant son.

"Now you will both die," he growled as they stumbled into the room.

"But, King," Jack whispered. "The ruby is repaired." Jack begged the king's permission to show his handiwork.

"But you've failed," came King Muenster's retort as he licked his spotted lips. "I can still see the crack."

Jack took the giant royal red ruby in his hands and turned it upside down. At the bottom, where the crack ran into the tip, Jack had intricately carved the petals of a rose. It was the most beautiful rose in the world. The spined crack had become a stem, traveling to the core. The king knew that the rare rose, carved into the equally priceless ruby, would be the most desirable piece of work ever to grace the walls of Fondue. He knew that his subjects would revel in his wisdom of turning a worthless cracked jewel into a beautiful flower.

King Muenster was so struck by his discovery of beauty that he ceased to be bothered by his scent. Pumpernickel spent the rest of her days soaking up the messes of Fondue. And Jack went on finding roses in her messes.

SECTION ⟨7⟩

If the People Want to Groove, Don't Make 'Em Polka

Beyond C-SPAN

*W*hen I signed up for the job of president of the Sierra Club, no one told me that every morning would begin before the *San Francisco Chronicle* landed on my doorstep. At 5:30 on a cold, December morning in Washington, D.C., I cabbed over to the C-SPAN studios for an appearance on *Washington Journal.* The show tries to be a bit more constructive than most Washington talk shows, rising a step above *Crossfire, The McLaughlin Group, Firing Line,* and *The Capitol Gang.*

Most Washington talk shows are little more than mental dildonics, showcasing cranky people who prove America's assumption that politicians are more concerned about ego and arguments than the fate of the nation. While disagreement is part of a healthy democracy, let's leave the wrestling to Jimmy "Super Fly" Snooka and the other tights-wearing macho men of the World Wrestling Federation.

The rest of the programming on C-SPAN seems to have the goal of putting to sleep all of the speed addicts in our country. From dreary-eyed debates between uninterested office staffers to word-by-word analyses of public policy legalese, the station appears to actively avoid covering anything of interest to those whose signature doesn't function as a postage stamp.

The typical morning at *Washington Journal* begins with a Styrofoam cup of instant coffee and five different newspapers. In fifteen minutes, guests scan the papers and select five articles of interest to discuss. The morning I appeared on the show I was joined by Ron Christie, a legislative aide to Representative John Kasich (R-Ohio).

We chose our articles, went into makeup, situated ourselves in the studio, and the cameras rolled. I started right in.

"Good morning. Let me start out by saying that Mr. Christie and I agree on ninety percent of everything. So if you're expecting a partisan brawl, you'd better change the channel." The host looked across the studio at his producer, who shrugged and wondered where the "guest formula" went wrong. "Oh well," the host muttered quietly, "it's early in the morning."

The first article I chose talked about the subsidization of logging roads. Under this U.S. Forest Service timber program, roads are constructed in forests to help private logging companies get into them for clear-cutting. More than 370,000 miles of roads, more than eight times the length of the interstate highway system, have been built on national forest lands. The annual cost of this subsidy is approximately $50 million. Timber roads are a typical example of how the government pays for companies to destroy our environment. Ron Christie agreed.

The producer began to shift uncomfortably, nervously looking over his shoulder for the invisible Nielson hand to crush him at any moment. Even C-SPAN can sense the coming of a "Bob Dole" moment, when the audience collectively falls asleep.

I flipped through the paper quickly, skimming the Business and Nation sections, and finally arriving at the Living Arts section. I held up the article that caught my eye. The headline screamed, SEINFELD SELLS SUBSIDIARY RIGHTS FOR $5,000,000 AN EPISODE.

"What does this mean to you?" I asked Mr. Christie.

"No idea," was his quick reply through a look of wary befuddlement. Apparently, Representative Kasich had no formal policy on *Seinfeld*.

I turned to the host, asking again, "Do you know why this is important?"

The host frowned. "No, but I'm hoping that you're going to explain it to us."

"There's a sad truth here," I said. "People care about *Seinfeld* more than they care about you. They're more interested in the humor of life in a New York apartment than in the quibbling of Washington's elite." I paused. "But it doesn't really matter what I say, because no one's watching anyway."

I was wrong. Someone was watching. When I got back to my office, I received a call from an irate C-SPAN fan. "You kid," he fumed, "you're an ignorant, insolent, precocious, left-wing, pinky, hippie, enviro-freak! What do you think about that?"

"Wow," I paused a second to absorb it all, "you learned all that just from seeing me for an hour on C-SPAN? You're pretty perceptive."

Reaching people through their interests is not a groundbreaking idea. David Brower created the activist coffee-table book, *This Is the American Earth,* with photographs by Ansel Adams and text by Nancy Newhall. The book was published to expose the American public to its landscape and to do so in a format that anyone could embrace. Reviewers called it "one of the great statements in the history of conservation" and credited it with delivering "the most important message of this century." It made a political and social message accessible through fine art and introduced the Sierra Club to a wide, new audience.

The books brought wildness into people's living rooms. They spread the message to reaches of the country never before touched by the Sierra Club. Not everyone could travel to Yosemite to appreciate its beauty, so Brower brought its beauty home. Our mission remains the same today—to reach people through their interests and to communicate to them using a language they understand. Timing is everything. We're always looking for the right moment to strike. We're prepared to strike back as soon as those who would destroy the Earth make their move.

Oil and Water

In 1990 an oil tanker, the *Exxon Valdez,* ran aground in Alaska. A velvety black blanket spread over the waters of Prince William Sound as eleven million gallons of crude oil spilled into the ocean. An estimated fifteen hundred miles of shoreline were oiled, including the banks of three national parks, five state parks, four state critical habitat areas, and one state game sanctuary. Oil seeped over Bligh Reef in Prince William Sound, washed up to shore, and coated everything in sight.

From harbor seals to the marbled murrelet, wildlife felt the sting. Thirteen percent of the seal population in the area was destroyed. Twenty eight thousand sea otters were killed. Endangered already by the encroaching practices of civilization, the marbled murrelet lost another 7 percent of its population in the spill area.

Exxon paid billions of dollars in cleanup costs and millions more in compensation claims. Commercial fishing in the area faced a four-year moratorium while the fish struggled to repopulate the waters of Prince William Sound. Fifteen native villages that rely for subsistence on resources such as shellfish and birds found their carefully managed natural grocery and clothing depots totally destroyed.

Imagine going into the Gap, the only clothing store in town, and finding everything drenched in oil. And imagine finding that the Stop & Shop down the street, your only source of Cheerios and Snapple, suffered the same fate.

With the extensive media coverage of the *Valdez* spill came public demands that Congress spring to action. Congress immediately passed regulations requiring that oil companies use double-hulled tankers to prevent similar catastrophes in the future. Responding to public outcry, businesses adopted the "Valdez" principles—a new model for corporate planning and responsibility.

The *Exxon Valdez* spilled its cargo of oil in a remote region of our country. Few people directly felt the event, sitting safely in front of their television screens in middle America. Reporters flocked from great distances to Alaska to photograph the damage, interview key participants, and report back to the American public. The event garnered national attention and inspired a generation, even though few would ever personally see the oil-soaked shores of Prince William Sound.

In 1993 a mysterious bacteria known as crypto-sporidium made its way into the water system of Milwaukee. In just two weeks it enfeebled the city's population, halted commerce, shut down schools, and closed city government. More than 400,000 people felt the effects of cryptosporidium as it leached into their bodies through the water they use to drink, bathe, and brush their teeth. It was a Michael Crichton novel brought to life.

Recognized as a pathogen since 1976, cryptosporidium lives in more than 65 percent of the surface waters tested in the United States. Twenty-three million Americans live in communities with no filtration systems for surface waters. Residents of major cities like Boston, New York, and Seattle share this terrifying bond. All water-borne outbreaks of cryptosporidium have occurred in communities where water utilities meet state and federal water standards with adequate filtration systems, which makes you wonder if our drinking water standards are sufficient. Cryptosporidium threatens millions of Americans who don't even know it exists.

When cryptosporidium infected Milwaukee, mothers had no other choice than to believe that their children genuinely needed to stay home from school. Kids weren't faking to get a day in front of the TV. Sons and daughters of aging parents missed work as they cared for their loved ones suffering the ills of diarrhea, painful gas, and nausea. Those who were already sick felt the added strain of intestinal disease.

Hospitals filled to capacity with people hit hardest by the bacteria, weakened or pained beyond the help of home remedies. Tensions mounted, hospitals overflowed, and people were left to contemplate how and why their environment was attacking them. Cryptosporidium, an invisible enemy, killed 104 people. Most of those killed were infants, people whose immune systems were compromised by HIV, and the elderly. The weakest members of society felt the harshest effects.

In the face of a widespread disease right in the heartland of our country, you might expect extraordinary media coverage, action by environmentalists, and a response from a concerned political system. When it was discovered that the culprit was lax clean water standards, you might expect an immediate reaction from Congress.

Instead of answereing the call for cleaner water, the next Congress led an all-out assault on the Clean Water Act and the Safe Drinking Water Act, attempting to weaken their foundations. Representative Tom Coburn (R-Oklahoma), a physician, addressed the problem from the floor of the House of Representatives. He said that cryptosporidium can be quite useful in helping us to identify people with compromised immune systems. Tell that to the families who lost their babies in Milwaukee.

HR 961, legislation weakening standards for the release of toxins into rivers, lakes, and streams and relaxing standards for mandatory cleanup, was approved by the House of Representatives in May 1995. One of the greatest environmental tragedies of the decade did not garner a response.

Although the *Valdez* oil spill occurred far away, immediate media response to the event made it a prime target for public concern. Certainly, birds foundering on tainted shores and oceans slick with oil made for spectacular images. The media and environmentalists spread the word quickly, people sprang to action, and our political system was forced to prevent repeats of the same accident.

In the case of the cryptosporidium outbreak, a direct attack on children, the elderly, and people with AIDS that occurred in the heartland, the media was lax in communicating the message that without strict standards, the water on which we depend for life could kill us. Congress went on slashing protections in the name of regulatory reform and in the absence of the call of conscience. If we are not prepared to respond to all offenses to our environment and to our health, we cannot successfully communicate the need for action.

"Report the Man Who Spreads Pessimistic Stories"

By the time we hear about events like the *Exxon Valdez* spill or the Milwaukee cryptosporidium outbreak, corporations and regulatory agencies have already used sophisticated public relations machines to communicate their versions of the facts. The only way to counter this is to understand their techniques and beat them at their own game.

Tactical public relations campaigns were born in 1914 when Ivy Lee, the son of a conservative Georgia minister, was hired by the Rockefellers to cover up the Ludlow Massacre. After hundreds of miners went on strike to protest inhumane working conditions, the Colorado Iron and Fuel Company sent in a militia to attack miners and their families. Some were beaten to death, some were burned. Fourteen people died at the hands of the militia.

Ivy Lee quickly earned the nickname "Poison Ivy" by launching a massive campaign to counter negative media coverage of the massacre. He worked quickly and steadily, sending out bulletins every four days to refute any negative information printed by others. He counterattacked, exaggerating salaries of union organizers and assaulting the character of labor organizers. He called Mother Jones, an eighty-two-year-old much-loved champion of unions, a "prostitute and a keeper of a house of prostitution." He created fictitious people to push forward his arguments and blamed the massacre on an overturned stove. He set a precedent for the corporate public relations campaigns of the future.

In 1917, one week after the United States declared war on Germany, President Woodrow Wilson furthered the idea of the "propaganda machine" when he established the U.S. Committee on Public Information (CPI). In an isolationist era, the CPI took on the public's fears that America's large immigrant population would create an insurgence at home.

The CPI commissioned seventy-five thousand "four-minute men," volunteers who stood up in movie theaters to give a four-minute pitch about the importance of the war. The four-minute men enlisted the help of citizens to watch their potentially traitorous neighbors with canned speeches like this:

> We must remain alert. We must listen carefully to the questions our neighbors are asking, and we must ask ourselves whether these questions could be subverting the security of our young men overseas. . . . Report the man who spreads pessimistic stories, or who asks misleading questions, or who belittles our efforts to win the war. Send the names of such persons— even if they are in uniform—to the Department of Justice in Washington.

George Creel, the civilian head of the CPI, liked to say, "People do not live by bread alone, they live mostly by catchphrases." Public relations campaigns had sold products. Now they took on a more insidious form—selling war. By removing the focus from the war and placing it on the imminent threat of your neighbor next door, the CPI used paranoia as a tool of enlistment. It involved the people it tried to reach and incorporated those people in the message. It gave them a catchphrase by which to act and live.

George Creel's idea has never been more true. Slogans like Nike's "Just Do It" have been co-opted for public use and appear on T-shirts to promote sex and surfing. Wendy's slogan "Where's the beef?"

became a common euphemism for questioning the content of a Reagan speech or the size of a body. Everyone I know is sick of people echoing the cry "Show me the money!" coined by the movie *Jerry Maguire*. It's time to turn these techniques back on the people who invented them.

Gustave LeBon, a German psychologist who provided many of the intellectual underpinnings of public relations, stated, "Crowds have always undergone the influence of illusions. Whoever can supply them with illusions is easily their master. Whoever attempts to destroy their illusions is always their victim."

Before you object to the notion of employing these brutal, misleading tactics to get a positive message across, let me point out a major difference. Environmentalists differ from private companies in that there is no direct self-interest in our message. I won't get rich if Congress decides to protect the Boundary Waters of Minnesota. We only need to provide people with the facts that enable them to make up their own minds and empower them to act on their decisions. We need to be savvy, but we don't need to employ scare tactics and misleading information to communicate our message.

People are tired of hearing bad news about their world. Many Americans have become jaded by environmentalists who run around like Chicken Little, predicting the end of the world as the sky falls down. While we have a responsibility to explain the problems our environment faces, constantly bemoaning the state of the world does not inspire people. Environmentalists must show how one person can make a difference. People do because they're told they can.

The environmental message is content-rich and signal-poor. We have great substance to our message—hope, triumph over adversity, the future—but we don't bother to signal it. A good idea that sits in the minds of a few smart people is wasted. Some environmentalists will claim that they have been signaling, but that the audience isn't listening. These environmentalists are profoundly arrogant in their communications techniques. They assume that everyone understands the "truth" just as they do.

Ed Mahe, a Republican political consultant, asks audiences a simple question to begin a speech.

"How many people here have had a specialty coffee drink in the past year?" he begins. "Lattes, cappuccinos, espressos, any snooty Italian coffee—how many people have had coffee other than your straight Bunn-o-matic drip coffee?"

Most of the people raise their hands in the typical political audience that he reaches.

Mahe explains that less than 2 percent of Americans had a specialty coffee drink last year. If you're one of them, you are in a tiny minority despite the fact that you wait in line for five minutes behind countless others for your morning latte.

Mahe's point is that your message need not necessarily appeal to you, as you may not be representative of your audience. The message is for the person you're reaching; it's not there for you.

River guides like Kathy Crist knew she had to make their message understandable. They worked and played on the American River near Sacramento, California, and faced a dam that threatened to put her out of business. The Auburn Dam was a billion-dollar pork barrel project that many scientists believe would never have worked to control flooding. Kathy wasn't going to stand for the loss of her liveli-hood and her favorite rapids. The river guides pulled together an otherwise fractious group of rafters to work for a common cause.

They set out a plan to send a message to Congress. They knew how beautiful and inspiring the American River is, but they knew their audience of politicians, whose idea of roughing it is driving a Range Rover through a puddle, didn't get the message. They orga-nized the river guides to collect a flood of letters.

For months the guides gave visitors a taste of the best ride the American River has to offer, bumping them through the class-four rapids of "Meat Grinder" and washing them with "Satan's Breath." When they reached the calmer flat waters of the river, the guides took advantage of their captive audience.

"Isn't this fantastic?" the guides asked. The tired, wet, and thrilled rafters responded with unanimous approval.

"Can you imagine that anyone would want to ruin it?" was the next question. The rafters, having felt the thunder of Satan's Breath, could barely imagine it.

The guides then pulled paper and pencils from their dry bags and had the rafters write letters on the spot. The efforts of Kathy and the team of river guides made members of Congress hear a message thousands of people strong. The Auburn Dam project was abandoned, in part because Kathy helped them feel the importance of the river using a signal they understood.

People Do (Mislead)

Industries take great care to tell stories that fit their interests. Chevron splatter-casts an ad campaign highlighting the proenvironment efforts of its employees. In one ad they show a little fox being eyed by a coyote in the distance. The text reads:

> Across the twilight of a California desert, a kit fox hears the deadly footfalls of a coyote. Caught in the dangerous open, she can streak for safety to a curious mound on the edge of an oilfield. People who work there, consulting with wildlife experts, built it especially for her.
>
> So now she can squeeze through a pipe just big enough for her and into a cozy den that's designed to keep her snug and safe.
>
> Do people think of things like this just to help an endangered species make it through the night? People do.

Chevron sends out a simple message of success in caring for the environment, greenwashing the image of an oil company that is one of the biggest polluters in the United States. In 1987, Chevron was fined $145 million for its environmental violations. It's ad doesn't mention that the real reason "People Do" care is that the Sierra Club sued Chevron to stop its destructive practices.

Sting

Like the simple message "People Do," a simple name can carry greater meaning. Gordon Sumner, better known as Sting, was once asked why he chose to name his band the Police. He paused, and in classic British style looked patronizingly at the reporter and asked, "Why do you think?"

The reporter was startled to find himself at the other end of a question. He collected himself and came back with another question, looking for quotes to fill his article. "Were you making a statement about the growing influence of authoritarian institutions in our society?"

"Guess again."

The reporter thought for a second this time. "Were you doing it to poke fun at the police who always seem to be attacking musicians for their fondness for drugs?"

Sting laughed at a man who was obviously trying

too hard. "I did it for one simple reason—all the free publicity. Every time you see a police car, it's a free advertisement for our band."

The most powerful messages utilize existing icons that are already rich with meaning. The Red Cross realized it. Andy Warhol took it to the extreme by celebrating the Campbell's soup can. By the end of his career, Warhol was being asked to autograph cans of soup. The Sierra Club links many of its campaign materials to a graphic of the American flag. Symbols move people and they are ours to employ.

What Roseanne Forgot to Tell You

Like it or not, television is America's most influential teacher. We can spend our time fighting it, or we can learn from its appeal. My friend Isaac Peace Hazard owns an early eighties Panasonic Compufocus set with a snap-in remote control. On the front of the set, six buttons provide direct and astonishingly accurate access to the world of television. The television was handed down and across generations until it came to rest in his South of Market apartment in San Francisco.

On top of the Compufocus rests a fairly modern set of rabbit ears with the added bonus of a circular antenna. The rabbit ears don't do much good because the cable-ready jack hangs in shreds at the back of the TV. Although the antenna pulls all free and public signals into the room, it hits a roadblock when trying to deliver those signals to the apparatus meant to translate them.

Isaac, who mainly watches TV in the times at night when he doesn't feel like going out and can't read any more, reports that between 10 P.M. and midnight on any given night, the best and generally only watchable program on his TV is *Roseanne*.

At times *Roseanne* entertains with sharp wit. Who can't identify with a child firing quick retorts back at his mother's complaints? Who hasn't wanted to counter a comment from Mom like, "Take off those muddy shoes," by taking them off and placing them on the kitchen table saying, "There, they're off."

But how is this useful? Which new doors of the mind open when we listen to a punk outwit his mom with tactics anyone has thought of and must suppress because of common civility?

A study by TV-Free America found that "[T]he average American watches more than four hours of TV each day. At this rate, by age sixty-five, that person will have spent nine years of his life watching TV." For children, the study goes on to reveal that "[U]pon graduation from high school, the average American child will have spent more time watching TV than in school."

Where is the upside in all of this? What is the positive factor in children receiving more education from the boob tube than they do from trained teachers? TV can be transformed into our ally. Television serves as the most universal form of communication to younger generations and offers a powerful opportunity for organizing.

Instead of bashing kids for watching MTV, let's use MTV to our advantage.

We can offer musicians with environmental or other socially beneficial agendas an opportunity to showcase their music and ideas about improving our world. Josh Sage created a thirty-minute TV special to run on MTV for Earth Day. When he asked me to participate, I remember thinking he was crazy for trying to produce the spot with virtually no budget and an impossible deadline. I was wrong. The production was a smash, with Michael Stipe from R.E.M. finishing the spot by saying, "Now, turn off your TV, and go outside." Thousands of young people surfed over to the show's website and signed up to get involved.

At this time, the point of at least one-fifth of all TV is to get you to consume—the magic box even features thirty-minute infomercials for hair plugs and the Psychic Friends Network. And consumption irrefutably leads to pollution, waste, and the exploitation of our natural resources.

Commercials tell you that you sin when you fail to buy a particular product. If your floor doesn't sparkle and shine, if you are shackled with ring around the collar, if your pizza doesn't have a stuffed crust, you will be looked down upon by those who have seen "the way," those who purchased the better product.

And heaven is just around the corner at the store. All you need to do is buy the bright red plastic bottle and the ring around your neck will fall away, leaving you to ascend into the circle of acceptability, tranquillity, and joy shared by others. The most unhappy people I know have more stuff than they will ever need or use.

The only heaven I can be sure of is the one I'll find here and now. And I'll still be able to find it even if my dishes aren't squeaky clean and I'm driving a rusty 1972 Ford Pinto. We can't value our lives in terms of what we own.

That said, the heaven I speak of includes televisions. Go ahead, call me a vidiot. Why ignore such a potent tool for communication? Why not speak to today's youth using a medium they understand? Speak *to* them, not *at* them. Appeal to emotions and intellect and you'll get them to turn off the TV occasionally.

Television is a double-edged sword. One side aggressively hacks at the viewer's desire to act by replacing experience and emotion with the drama of a two-dimensional reality. The other side is poised, although frequently idle, waiting to cut the very ties that bind the user to it by causing that viewer to think and to act upon those thoughts.

Isaac once likened TV to cotton candy, saying, "It tastes good going down, so you greedily pick at it and blindly stuff it in your face. Then you realize it made a total mess of you, it's stuck to you in places it never should have touched, and you feel sick because of it. And although you eat it for hours, at the end you still have no energy."

TV could be protein and carbohydrates. Sometimes it is. The key to TV as a tool for communication resembles the key to good nutrition. Put nutritious programs, rich with energy, on the table. Throw in the occasional sweets to balance the occasionally necessary brussels sprouts and lima beans. Encourage a TV diet every now and then. At least be conscious of what you cook up and why, and stay tuned to the effects of what you serve every day.

Wired

Peter Bahouth knows what it takes to get the media's attention. He's used just about every available technique of communication he can to reach the American public. As the executive director of Greenpeace, he wrote the book on visual activism. Upon graduating from the New England School of Law, he opened a private practice in Boston to represent evicted tenants, earning a grand total of $45 in his first few months. He didn't need money to be happy, but he still wasn't fulfilled.

Peter's law office sat above a restaurant where Greenpeace staffers ate lunch. He had seen Greenpeace on television and thought, "Here's a group who sees when something is wrong, they react, they are active, they break people's perceptions about what you're supposed to do when confronted with injustice. And it's so damn visual . . ."

As a child he was quiet, living mostly in his head and observing rather than interacting with his surroundings. He grew up in New York on Onondonga Lake, the most polluted lake in the country. "It was the mid-fifties," he remembers, "and there wasn't even the word *environmental*. But I knew that it was wrong that I couldn't swim in the lake. I remember seeing the company dumping its wastes right in the water. I knew that was wrong." Peter watched as a native species of whitefish became extinct from chemicals illegally dumped in its spawning ground.

The activism of Greenpeace, bold in its content and imagery and broadcast to millions, attracted Peter and moved him to volunteer his time as a lawyer. He represented Greenpeace in civil disobedience cases and once even helped out when Greenpeace sued itself. Greenpeace started in Canada with a chaotic plan to go forward and multiply throughout the world. When the group operating under the Greenpeace name in San Francisco decided they were independent and didn't need to send money to Greenpeace Canada, the parent organization sued the rebellious child. The end result was the formation of Greenpeace International, a coalition of independent, national organizations.

Peter was asked to sit on the board of the Boston chapter. He knew right away that the time was right for Greenpeace to make a move. "The mixture of visual and vision was ripe," he says of the activist tactics characteristic of Greenpeace. "I suggested that we come together to form Greenpeace USA. I worked with each one of the eight smaller groups in the U.S. to bring them together." It was no easy task. Greenpeace members were rebels to the core and didn't trust structure.

In 1988, Peter was appointed as executive director of Greenpeace USA. He called the new position "just one more step in on-the-job training." At the time, he didn't particularly want the position, adding a new twist to the Groucho Marx saying, "I wouldn't want to belong to any club that would accept me as a member." His reluctance made him more attractive to the organization.

Under Peter's direction, Greenpeace aimed at appealing to instinct over intellect. Based on the Quaker principle of bearing witness and the non-violent civil disobedience tactics of Mahatma Gandhi and Martin Luther King, Jr., Greenpeace sought to reach people's hearts by getting out there and getting in the way. "We weren't scientists, we were communicators," says Peter. "Here was a group whose role was to take the high road and stand for the environment, not to be a part of the junior politicians club of Washington. People knew where we stood. We weren't vague."

On the twentieth anniversary of Earth Day in 1990, Peter chained himself to Du Pont's tracks to stop the shipment of chlorofluorocarbons (CFCs). CFCs are the primary depleters of the ozone layer, and Du Pont is their primary producer. Although Du Pont has committed to phasing them out, their replacement is HCFCs, an only slightly less damaging product. It takes about fifteen years for ozone-depleting chemicals to reach the ozone layer, where they remain active doing their dirty work for years. If anyone doubts the seriousness of the destruction of the ozone layer, ask an Australian. Almost every Australian over the age of seventy-five has some form of skin cancer.

Peter's actions sent a clear message to Du Pont and provided a compelling image to the media. "Our statement was," says Peter of the event, "we will stop the trains from going out. We will stop the shipping of CFCs. I chained myself to the tracks and ground them to a halt."

Greenpeace members place themselves in the paths of environmental offenders, providing the front line of defense for our world. They wedge themselves between Japanese whaling ships and surfacing whales in small rubber Zodiac boats. They step in between seals and their assailants, risking their health and even their lives. And true to Peter's style, they always get it on tape so that the message can be shared with a greater audience.

Today, Peter is the executive director of the Turner Foundation, founded by environmentalist and communications magnate Ted Turner and his activist wife, Jane Fonda. The foundation gives millions of dollars to environmental causes, mainly funding small grassroots organizations.

As executive director of the Turner Foundation, Peter spends more time in offices than in controversy. But he still adheres to the principle of getting out into the field and understanding firsthand what is happening. The foundation places a high priority on site visits; Jane Fonda and Peter try to visit every group to which the foundation makes a grant. The foundation is his headquarters. The projects are his mobile warriors.

In a conversation, I asked Peter, "Why do people become activists like you?"

His response: "Everyone's got the wiring; in some people it's just more exposed."

SECTION 8

A Block Party in Every
Neighborhood

The Concrete Jungle

*G*rowing up in a rapidly suburbanizing town on the outskirts of Los Angeles, I witnessed my community being infiltrated by bland, Kmart culture. When I was younger, orange orchards and horse farms lined the streets. I remember kids riding horses down Vanalden Boulevard over the last dirt roads that connected the different neighborhoods. We enjoyed meals at Cables, a local diner where we knew the waitresses. None of them wore name tags; they didn't need to.

My town was founded by Edgar Rice Burroughs. You might know him as the author of *Tarzan*. In a streak of brilliance, Borroughs named the town Tarzana. Tarzana always laughed at its identification with the Wild West, with big stagecoach wheels resting outside the rustic facade of the post office and a Mexican restaurant with sieves once used by miners to pan for gold.

As the suburbs of Los Angeles grew, horses were zoned out because they were inconsiderate enough to attract flies. The city paved over the last dirt roads so

Tarzanians could speed to wait in L.A. traffic. The gold pans gave way to the paper wrappers of Taco Bell.

Tarzana, the suburb of the suburbs, is ground zero for the takeover of community. We were born and bred for it.

Before long, you couldn't find a local diner within miles of Tarzana. Chain restaurants came on the heels of malls, the muscle of Burger King and Red Lobster pushing out the character of Cables. Three McDonald's restaurants were built within three miles of my house. Eventually, the town tore down the rustic post office and erected a mammoth pink-stucco postal center. They even junked the wagon wheels.

Our suburban next-door neighbors were distant. The only neighbor with whom we had contact tried to sue us for allegedly training our dog to shit on his lawn. He then threatened to sue us for encouraging our trees to grow into his yard.

Tarzana was home to the eighties icon of the Valley girl. I used to walk through the malls hearing the shrill cries of, "Like, oh, my *God,* gag me with a spoon!" I'm probably the first president of the Sierra Club to be a mall-cruising Valley boy. I'm not particularly proud of that. Identity in the Valley was a reflection of established trends rather than an expression of self.

Tarzana did provide me with a distinct sense of place when I explored the Santa Monica Mountains, the scenic backdrop to my house. About one mile up Reseda Boulevard the road petered out into chaparral forests. Manzanita bushes, sage, rosemary, and scrub oaks stood next to each other like children in a third-grade class picture. People used to call it an elfin forest because no plant reached higher than six feet. It was desert and it was beautiful. It was my sandbox.

Around 1990, a developer began to talk about continuing Reseda Boulevard right over the mountains and into Santa Monica. He mapped out one thousand $2 million home sites on the richer side of the mountain that offered spectacular ocean views. Tarzana, already a victim of fast-food chains and strip malls, became incensed at the idea that our mountains were to be altered. Apparently I wasn't the only person who found a sense of place in the mountains of our town.

We joined forces with the Sierra Club and the Santa Monica Mountains Conservancy to halt the development. I remember feeling proud of my fellow Tarzanians.

Nature provides a sense of place because it is always unique—no slope of mountain, stretch of desert, turn of river, or arch of stone is exactly like another. A Wal-Mart or Home Depot in Louisiana and Montana are essentially identical in both look and feel. The bayou and the plains are just as different as the Creole and the Cree.

Towns throughout the world are losing what Wendell Berry calls "a sense of place." In Lawrence, Kansas, a new Barnes & Noble opened and forced the three smaller independent booksellers out of business. New subdivisions and suburbs in the town are so large

that Lawrence needs a second high school. The suburbs, mostly white, will fill the new school with white
students while the inner-city school will become
almost entirely African-American—de facto segrega-
tion from urban sprawl.

The state of Maryland will consume as much land
in the next twenty-five years as it consumed over the
past three hundred years, more than 500,000 acres
of forests and farmlands. Culturally rich urban centers are being replaced by endless lines of faceless
boxes, differentiated only by the limited palette of
developers' colors. And amongst these cookie-cutter
suburbias, the tired corporate logos of chain stores
provide little variation.

Hospitals, meeting places, barber shops, and munic-
ipalities follow the people outward. As population centers lose their utility, civic values fade. Downtown will
no longer be a meeting place where neighbors greet
each other and businesses exchange services.

Seventy percent of Americans don't even know the
people next door. The only contact they have with
their neighbors is through silly lawsuits and furtive
glances. If the environmental movement were really
serious about success, it would throw a block party
in every neighborhood in America.

Wallace Stegner called the canyons and slickrock
of the Escalante "the geography of hope." We are
turning the rest of America into the geography of
simulacrum—a mere trace of itself, an insubstantial
representation of what once was and what could be.

When we know and understand the people in our communities, our communities will flourish. The sense of place will return, replacing the sprawling arms of a standardized and sanitized culture.

Fortunately, some towns have never lost their sense of place.

I've been told to go to Hell many times this year. My friend Caleb went there and gave me a report. I was happy to hear that United Airlines flew close by. For those of you who don't know, Hell is a town in Michigan. Caleb looked, but there's no church in Hell. There is a biker bar. It seems everyone wants to get tanked in Hell. The main attraction in Hell is an enormous United States Weather Service weather station. People call in from all over the country to the weather station in Hell when the temperature gets icy. Every TV weatherman in America wants to know when Hell freezes over.

There are no museums or promenades in Hell, Michigan, but the people who live there are proud of it. They once even debated changing the name of their town. It's not easy living in Hell. But the people who live there would never relieve themselves of the character of home.

A return to community is our only hope for saving the diversity of our country and the unique character of our environment.

The Pact

Radical localism offers us a chance to lessen our impact on our environment through a return to community. Before you get nervous about the word "radical," remember that it means getting to the root of the problem. The root of many of our environmental and social ills is a global trend away from local communities.

If you care about our environment, jobs, education, transportation, and quality of life, it's time for us to join in a pact. It's a pact that requires us to stretch ourselves. We can't be perfect, but we need to strive to support the most local product possible in order to support local communities and keep people connected to them.

Although a radically local community focuses on developing its own strengths and resources, it is not isolationist. Instead, it supports diversity in life and in people. It readily accepts people from outside communities who wish to bring their skills and ideas to better the community. Radical localism in no way serves as an excuse for fearful isolationism.

A radical local community adheres to a few basic tenets. To preserve its identity and its resources, and to preserve our environment, a radical local community sees natural resources as its nest egg and only spends the interest. Forests are selectively harvested and wilderness is preserved to maintain biological diversity.

The radical local community values personal interaction that leads to community building. Fewer E-mails are written as face-to-face conversations resurge. As my friend Andrew Lang says, "There must be less typing and more talking."

A radical local community values humans more than machines. If true costs are equal or close, companies give humans jobs rather than investing capital to buy a machine that will make a more perfect etching. Mahatma Gandhi once lamented, "It is a tragedy of the first magnitude that millions of people have ceased to use their hands as hands. If the craze for machinery methods continues it is highly likely that a time will come when we shall be so incapacitated and weak that we shall begin to curse ourselves for having forgotten the use of the living machines given to us by God."

Focusing on long-term quality and value, not short-term profit, gives the radical local community its future. Planning provides goals for the future and preserves the values of the community. The community holds ecology, the science of connections and interrelatedness, in higher esteem than chemistry. The answer to bugs on your tomatoes is not a new pesticide, but ladybugs. Humans have a tendency to find complicated solutions to simple problems, but the science of ecology suggests that the answers already exist in nature.

Radical localism aims to preserve ecological integrity, local culture, and local economy. People fight most savagely to protect their homes; it's time to give them the chance. The root of the word *ecology* is *ecos,* or home. Let's go back to our roots.

The Radical Localism Pact

I will do my best to:

1. Buy products produced locally, products made with local ingredients and local labor.

2. Demand that outside corporations respect local culture and incorporate local products into their product base.

3. Know my community, both human and wild.

Pact Point #1: Swadeshi

A crucial step in achieving a radical local community is relying on local resources. The question is not whether or not people in Vermont are entitled to kiwis, but whether or not their maple syrup should travel thousands of miles and bear the grinning face of Aunt Jemima when the world's finest maple syrup is produced locally. The goal of localization is not to eliminate trade, but to reduce unnecessary transport that incurs expense, pollutes our world, and removes resources from the local arena.

Buying Mrs. Smith's frozen apple pie in rural Massachusetts where apples grow wild makes no sense. The taste of a homemade pie bought from a roadside farmstand beats the preservative-flavored recipe of Mrs. Smith any day. Buying locally fuels the local economy, connects a community, and saves the costs incurred by shipping goods around the country and around the globe.

The average food item travels 1,300 miles before becoming a meal. Mongolia has twenty-five million cows and goats. Yet even with this resource right at home, Mongolians sell mainly German milk in their shops. In Kenya, butter from Holland sells at half the price of local butter. In England, the British can purchase butter from New Zealand at a much lower cost than the butter from their backyards. What's next? The Masai wearing the pointed clogs of the Dutch?

For each pound of food sent across our country, several pounds of petroleum are burned, consuming limited resources and contributing to global warming. The massive amount of oil used to ship goods far and wide also leads to costs related to protecting that oil, like the Persian Gulf War. And the trucks that burn the oil need roads to travel. In the United States we have 2.5 million miles of paved roads. $100 billion has been earmarked for highway development and maintenance over the next five years.

Twenty-five percent of the food grown on industrial farms spoils before it gets to the dinner table, wasting one-quarter of our soil's ability to provide us with the nutrients we need for life. Agricultural subsidies favor the waste of big-business farmers, providing biotechnology and tax breaks to those who produce massive quantities and use massive amounts of pesticides to do so. If you've ever tasted a fresh, organically grown, locally produced strawberry as opposed to one shipped hundreds of miles from Mexico, you know why local is better.

Similarly, large-scale energy production incurs large-scale environmental costs. Our government funds the development of pollutant-emitting coal plants, waste-dumping nuclear power plants, and ecosystem-threatening hydroelectric dams. Rather than focusing on locally based, renewable energy like power produced by the sun and the wind, we choose these grandiose methods of fueling our desires.

To move toward a focus on local strengths we must encourage what Mahatma Gandhi called *swadeshi*, which translates into "home economy" or "local self-sufficiency." According to *swadeshi*, whatever is produced in the village must be used, first and foremost, by the members of the village. Ghandi once said, "The true India is to be found not in its few cities but in its 700,000 villages. If the villages perish, India will perish, too."

You don't need to go to India to see *swadeshi* in practice. In Ithaca, New York, the Ithaca Hours system, established by local economist Paul Glover in 1991, utilizes community resources and keeps the fruits of village labor at home. In its first five years of existence, the program created and distributed nearly $50,000 in local currency. The currency has been distributed to more than nine hundred participants and is used by countless other citizens of the town. All of this contributes to what Paul Glover deems as Ithaca's GNP, "Grassroots National Product."

Ithaca Hours hold a value of ten dollars, the approximate average hourly wage in the area. From the two-hour notes to the eight-hour notes, the currency depicts native flora and fauna, pictures of the rural landscape, and sometimes the faces of locally respected community members. Some bills are printed on locally made paper, using cattail fibers from Ithaca's backyard.

Participants in the program use Ithaca Hours for everything from monthly rent payments and home repair expenses to medical services and child care. Two of the large locally owned grocery stores and farmer's markets in the area accept the bills as payment for the food they provide. Even the local Ben & Jerry's gets in on the game, accepting Ithaca Hours in return for syrupy sundaes, milkshakes, and other warm-weather treats.

The free newsletter *Ithaca Money* circulates six times each year with listings of member businesses and services, advertisements, and announcements. Participants pay one U.S. dollar to join the program, for which they receive four Ithaca Hours. Businesses listed in the newsletter apply every eight months to receive two Hours for continued participation. The currency is distributed carefully and gradually to avoid wild growth.

Glover explains, "With Ithaca Hours, everyone's honest hour of labor has the same dignity." He cites the necessity for some participants, like dentists who pay large sums of U.S. dollars for supplies, to charge more than one Hour for an hour's work. "Still, there are situations where an Hour for an hour doesn't work . . . so a lot of negotiating must take place."

The Hours are used as charitable donations to deserving community organizations. To date, more than $4,000 in Ithaca Hours has been donated to more than twenty local organizations. Loans of Hours are given without the burden of interest. "We regard Hours as real money," Glover explains of the community's sentiment, "backed by real people, time, skills, and tools." The system reflects another principle of Ghandi's *swadeshi* philosophy—not mass production but production by the masses.

Glover has shared his success with other local communities by creating a Hometown Money Starter Kit, which sells for twenty-five U.S. dollars or two and a half Ithaca Hours. The kit has made its way to four hundred different communities throughout forty-eight states. Already, printed local currencies are used by twenty-one communities from Eugene, Oregon, to Santa Fe, New Mexico, and across the country to Syracuse, New York. Local success translates into global success.

While Glover encourages the employment of similar systems, he warns people to know their communities before trying to implement a program. He suggests active outreach to garner support of small businesses and a diversity of services. To begin a system, one must know one's community, its resources, and its needs.

Pact Point #2: Cheap Underwear

With differing climates and the development of different technologies in different areas, no one community can be entirely self-sufficient in our country without a significant change in lifestyle. An attempt to grow wheat or rice in the sandy soil of Cape Cod teaches a quick lesson about the capabilities of the soil there. To tell L.A. that it has to produce all of its subsistence goods amid its highways and tightly packed buildings is absurd. Some products we consume will come from major distributors. But as part of our effort to create radically local communities, we need to hold major corporations responsible. We should demand that the Safeway in Idaho carry only native potatoes. And we should draw the line when department stores bottom out prices, muscle out local businesses, and eradicate local culture.

Al Norman, the Wal-Buster, was a hippie. "Absolutely," he chimes right in when asked, "I was drawn into politics fighting against the Vietnam War. The injustice was clear."

Al attended the University of Wisconsin and contemplated becoming a teacher after graduating. Instead of restricting himself to a classroom, he chose to share his messages with a larger audience. He worked for a while as a columnist and book reviewer for *Newsweek*. And then he started preaching to people about the quickly spreading evil of Wal-Mart.

Al Norman calls Greenfield, Massachusetts, home. Greenfield is a small town of eighteen thousand people in north-central Massachusetts, halfway between Boston and the mountains of Vermont. Characterized by open meadows, historic homes, and hard-working people, it is the epitome of early American frontier towns.

"I'm not saying Greenfield is some sort of paradise," Al says of his home, "but it's not a sacrifice zone either." Greenfield almost became a sacrifice zone when Wal-Mart swaggered into town, attempting to open a store. In October 1993, a friend asked Al for help in passing a referendum that would prevent Wal-Mart from building an enormous store. He remembers, "I started looking into what building a Wal-Mart would mean to Greenfield, and I began to become fascinated with how disgusting they were."

Al's community faced a new breed of toxin. The long pink aisles found in most Wal-Marts, stuffed with every gadget, accessory, and toy for an all-American Barbie doll, could wreak havoc on a town.

Al frowns on Wal-Mart's smiling image, which he believes covers up a business scheme intent on sucking life out of local communities. He objects to the trade of a small community's soul for the convenience of row upon row of imported, low-quality junk—anything you might need for your work, home, or pleasure.

Al organized citizens in his hometown to oppose the megastore invasion. Expressing public opposition to Wal-Mart's presence wasn't enough despite the statement by the company's founder, Sam Walton, that "we have almost adopted the position that if some community, for whatever reason, doesn't want us in there, we aren't interested in going in and creating a fuss." They must have a strange definition of "fuss" in Walton's small hometown in Arkansas.

Although this became the company's verbal message to concerned citizens, Wal-Mart's actions proved otherwise. In Greenfield, Al and his community group were forced to attack zoning laws to block Wal-Mart, which spent nearly $25,000 hiring a public relations firm to counter prevailing opinions in the town.

The idea that Wal-Mart willingly refrains from entering communities with strong opposition to its presence makes little sense to residents of Ticonderoga. When they opposed a store in their town, Wal-Mart bused in employees from nearby Plattsburg to heckle objectors. In Warwick, Pennsylvania, while town supervisors voted unanimously to shoot down a proposed 200,000-square-foot store, Wal-Mart refiled a new plan, called "basically a duplicate" by the town manager. It claims to have a community's interests in mind but repeatedly hammers to establish stores in the face of public opposition.

Al used the zoning laws of Greenfield to halt the invasion of Wal-Mart but didn't stop after winning the battle. He continues to fight the rest of the war.

Al Norman sends out a monthly *Sprawl Busters Alert,* an antimegastore newsletter. He travels across the country, helping grassroots groups organize against big-business invasions of their small-town cultures. In one edition of the *Sprawl Busters Alert,* Al recognizes, "There are now at least forty-five citizen's coalitions actively fighting megastores, and many of these coalitions have expressed the desire to ratchet our effort one level higher by combining forces towards a larger battle."

"There's one thing you can't buy at Wal-Mart," says Al. "That's the small-town quality of life. And once you lose it, you can't get it back at any price." Norman has been called the Paul Revere of the anti-superstore movement, igniting a torch and carrying it to other small communities across the country in an attempt to warn them of the imminent danger. The *Philadelphia Inquirer* calls him the "Wal-Mart nemesis."

Al fights Wal-Marts for two reasons. "One is bad economics. The second is quality of life. There's more to life than a cheap pair of underwear."

New Wal-Marts tend to be built outside of established commercial districts in towns such as Greenfield. By locating their stores outside of developed areas, Wal-Mart uses up farmland and takes advantage of lower property taxes. While enjoying this economic benefit, the stores cause communities to extend roads and services to their remote locations, placing a strain on local budgets.

If Wal-Mart stores pumped money back into the local economies of towns, their impact wouldn't be so severe. But money spent at the megastores goes directly to the corporation, which uses profit to expand into more communities. University of Massachusetts researchers found that a dollar spent at a locally owned business has four to five times the economic spin-off of a dollar spent at Wal-Mart. Charitable donations from Wal-Mart's local stores amount to a mere .0004 percent of sales, according to anti-Wal-Mart forces in Lake Placid, New York. This figure translates into about $4,000 per store each year. In 1995, Wal-Mart's net sales were more than $82 billion.

Wal-Mart officials often point to the 622,000 people it employs as one of the economic benefits it provides. The jobs rarely pay above minimum wage. The average annual income for a full-time Wal-Mart worker is about $12,000, well below the poverty line. Maybe the company should add wages to its claim of being the low-price leader.

The large number of people Wal-Mart employs

doesn't account for the small-business owners they force out of business either. Because they buy merchandise in huge quantities to stock their megastores, Wal-Mart reduces prices to levels unreachable by smaller stores. In 1993, an Arkansas circuit court found the company guilty of selling merchandise below cost to drive smaller stores out of business. The court awarded $300,000 to three local pharmacies as restitution.

Glenn Falgoust, a former store owner in Donaldsonville, Louisiana, explained the problem on a 1995 edition of *60 Minutes.* "In the ten years before Wal-Mart opened, we had a total of twenty business failures. In the ten years after Wal-Mart was here, we had 185 business failures. You could buy a bicycle in eight locations in this town. Today, if you want to buy a bicycle, you can only buy it at Wal-Mart."

The store once owned by Falgoust and his wife sold general merchandise, from bicycles to toys to lawn mowers. Their business was thriving in their eight-thousand-person town until Wal-Mart arrived. "Our business basically shut down in '86, after three years," explains Falgoust. "We stayed open because we had savings and we were young, and like [my wife] Angela tells me, I was totally dumb."

According to a study conducted by Kenneth Stone, a professor of economics at Iowa State University, retail stores within a twenty-mile radius of Wal-Marts suffer an average 19 percent drop in sales in the five years after the megastore invasion. Ten years after Wal-Mart entered the state, half of Iowa's men's and boy's clothing stores and a third of its hardware and grocery stores went out of business. Not only the businesses in "Wal-Mart towns" suffer, but also businesses in surrounding communities.

A decade after Wal-Mart entered Glenn Falgoust's town and sucked away his livelihood, downtown Donaldsonville is mostly dead. The new jobs created by Wal-Mart have been more than offset by business closings. As more and more people became unemployed, the tax base in Donaldsonville eroded, services declined, and the middle class fled. Falgoust laments the loss of his community and its icons, saying, "When you see our old general store that's been open for a hundred and seventy years, that survived the Great Depression, and it closes because of Wal-Mart? I mean, what—where does fair begin here? When do Mom and Pop get rights?"

Wal-Mart's first store opened in 1962, a small store in a small Arkansas town. Its baseball-cap-wearing, big-grinned founder, Sam Walton, took out a $20,000 loan to open his store in true American-dream fashion. It seems the American dream changed in the eyes of the company as it spread throughout the country like a virus, infecting and destroying American culture.

In Silverthorn, Colorado, Wal-Mart is trying to dynamite a mountain, to remove part of the natural landscape that distinguishes the state, in order to build a new store. One citizen placed a small sign on the mountain that simply read SAVE ME.

In February 1996, Wal-Mart announced plans to build a 93,000-square-foot store on George Washington's boyhood farm. Cessie Howell, a Virginian leading the "Save the Washington Farm" coalition, appeared on *The Today Show* on Washington's birthday to reveal this affront to patriotism. She wondered how anyone could want to replace our first president's farm with a concrete and neon megastore. She suggested we draw the line for profit around the boyhood home of the father of our country.

Are we willing to replace the roots of George Washington's cherry tree with the foundation of a Wal-Mart? Speaking to more than 250 residents of Stafford County, Virginia, where the farm is located, Al Norman brushed the snow off his shoulders and looked out over a crowd bundled in scarves and dressed in Revolutionary War costume.

"I cannot tell a lie," he began. "This is the most incompatible location for a Wal-Mart that I have *ever* seen in all my travels across this country. Combining Wal-Mart with Ferry Farm is the worst marriage since Lisa Marie and Michael." He held up a pair of white briefs, the blue-striped waistband stretched between his hands. "Wal-Mart is like a cheap pair of underwear. They keep creeping up on you."

A couple in the crowd wrapped themselves in an American flag and held a sign that read BY GEORGE, WE DON'T NEED ANOTHER WAL-MART. Less than five miles from where they rallied on Ferry Farm stood Wal-

Mart store number 1,833, in Fredricksburg, Virginia.

Wal-Mart comes to communities promising to enhance their quality of life. It offers all the goods and services you need under one roof, from tax preparation to beach chairs. It enters small towns, promising to boost economies and encourage growth, but profit quickly takes the driver's seat.

The town of Nowata, Oklahoma, with its 3,900 inhabitants, wrestled with the decision to embrace Wal-Mart in 1982, when sinking oil prices killed much of the profits of old oil wells in the area. Nowata pushed ahead and accepted Wal-Mart as a cultural and economic center of life, replacing the small businesses of downtown with "downtown Wal-Mart." People flocked to the shiny new store for its incredible variety of wares and low prices. Many local businesses went under, unable to compete with a store connected to a giant corporation. But everyone got to know Earl Duvall, the greeter who welcomed shoppers. A visit to the store became a visit with the neighborhood.

In 1995, Wal-Mart decided to close the doors of the Nowata store. A *New York Times* article reported Bryan Lee, president of First National Bank in Nowata, saying, "They were not playing fair. They came in and ravaged all the small businesses. And when it came to the point where they were not satisfied, they left." Armel Richardson, the mayor in 1982, remembers Wal-Mart's entrance into Nowata: "They said, 'We're here,' and we welcomed them, but Wal-Mart has proven this—they're big and greedy. They have no compassion for the community or the individual."

Wal-Mart pulled out of Nowata without explanation, saying it couldn't discuss the matter for competitive reasons. According to the bank president, Mr. Lee, financial losses couldn't have been responsible. The sales receipts left at the bank remained strong and the store's sales tax payments didn't falter either. Wal-Mart's departure was linked to the superstore it operated thirty miles away in Bartlesville, Oklahoma. A superstore's goal is to draw consumers from an entire region, and smaller stores, like those in Nowata, defeat that purpose.

Each resident of Nowata felt the economic strain of Wal-Mart's departure. The loss of the 3 percent tax on Nowata's biggest business's sales sent the town's budget plunging almost $80,000 into the red. As a result, the town laid off employees, abandoned repair projects, and raised sewer taxes 32 percent, and City Hall closed its doors an hour early each day. The city manager expected to ask for a five-dollar monthly tax on homeowners to provide the basic service of fire protection.

Wal-Mart entered Nowata with promises it had no intentions of keeping. Shortly before it closed its doors, the Wal-Mart in Nowata erected signs proclaiming, THE RUMORS ARE FALSE: WAL-MART WILL BE HERE ALWAYS. When it pulled out, Wal-Mart left Nowata with a cheap offshoot of itself. Bud's, a store that sells damaged and discontinued goods, came to fill the hole that Wal-Mart left. Red and yellow signs inside Bud's advertise, CLOSEOUTS. OVERRUNS. NAME BRANDS AT THE LOWEST PRICES! Sam Walton peddles his damaged goods under the name of his brother, Bud. Apparently, Sam and Bud didn't get along very well as kids.

Shoppers found the selection of Bud's goods to be an inadequate replacement. Kay McKee, a mother of two, explains with a bit of disgust, "I came here for a picture frame and shoes for her," referring to her sixteen-month-old daughter. "I wanted a twenty-by-seventeen frame. But frames only went to eleven-by-fourteen. Shoes, they had just boys', no little girls'. It makes me angry. Move all the little stores out, then leave. We're stuck."

Towns like Nowata experience the "Wal-Mart quality of life" and wonder where their culture went. Glenn Falgoust, the small-business man from Donaldsonville, Louisiana, explains how Wal-Mart affects quality of life, saying, "Today, the whole downtown section of our town, where people could actually walk to, the disadvantaged who didn't own cars, that's all shut down. Now they have to walk three miles down the Bayou Road to get to the Wal-Mart store. There's no sidewalks, there's no public transportation."

That's why Al Norman spreads the word to local communities across the country and helps them fight the invasion of their homes and their lives. "I wish I had more time," Al shrugs. "It's tough doing this work along with my day job." He pauses a brief moment and continues, "If people just realize that corporations can't just roll into town and do whatever they please, we're really going to get somewhere."

Pact Point #3: Keeping Track

In order to protect our communities from the onslaught of Wal-Mart and other threats, we need to know our communities. From the people next door to the land that surrounds us, our communities offer themselves as living records of time. All we need to do to is knock on our neighbors' doors to better understand why our communities function the way they do. Exposing ourselves to the wild areas around us will help us understand the need to protect those areas and the beauty they offer.

When people lived off the land they were familiar with its bounty and challenges. In today's "modern" life we need help to understand the place we call home. It can even be fun. Enter Keeping Track.

In Jericho, Vermont, hills spread themselves around the expanse of Wolf Run. Somewhere a catamount, just a spot in this colorful pointillist picture, winds its serpentine way through the underbrush of ferns, blueberry bushes, and sapling maples. Wolf Run is a two-hundred-acre piece of privately owned land nestled in the heart of the Green Mountains. Two hundred square miles of undisturbed wilderness, the largest slice of land in Vermont able to make this claim, surround it and support a diversity of life unparalleled in New England. Wolf Run offers the comforts of home to black bears, catamounts, deer, moose, and Sue Morse.

On a spring day, Sue had visitors. Students and teachers from a small Catholic school just north of Burlington arrived to learn about the wildlife of the pristine haven and to learn how Sue reads the signs left by the wild creatures of Vermont. Nuns in habits, young girls in pigtails, and boys in jeans permanently marked by ground-in dirt gathered around Sue as she prepared them for a trek through the woods.

One thirteen-year-old boy in particular stood out. He was a throwback to the fifties. His dark hair stuck to his forehead, a cowlick slicked down by the greasy adhesive of hair cream. His jeans, a bit too short at the cuff, revealed worn canvas tennis shoes. Despite his young age, Matt looked as if he should have a pack of Camels rolled up in the sleeve of his tight white T-shirt.

The band of nuns and students headed into the forest. Shortly after stepping out of the grassy, flower-spotted fields and into the shade of birches, firs, and maples, the group rounded a corner. Their mission was to read the signs of nature, and not thirty minutes into the journey they stumbled on a bent balsa fir sapling. It stood at a height of about ten feet and a circumference of about eight inches.

Its needles lay on the ground like a mess of brown and tan pick-up sticks. Its trunk pointed perpendicular to the ground, while the roots at the base fingered the air, telling of a strong force that blew through and uprooted the young tree.

Next to the sapling stood a larger tree, one that reached maturity with an upright posture. The low branches on the tree dangled like the inert arms of an unattended puppet. If the group had been looking for sticks to roast marshmallows over an open campfire, they would have been satisfied by this cache of brittle limbs. Thrash marks covered the tree seven or eight feet up, where brittle bark shavings hummed in the light breeze. To the untrained eye, the scene didn't look unusual. It was all just evidence of the natural process of death and rebirth shown in so many ways in the depths of wilderness.

To Sue's eye the scene told a story. The sapling and its neighboring tree offered a tale more descriptive than the writings of Shakespeare, Wordsworth, or Frost. And she brought this unlikely band of trackers here to help them unravel the yarn laid out by nature.

Sue walked up to the sapling and poked at it, examining it as if it were the first time she'd seen such a sight. A few brave students followed close behind her to get a better look. Anxious to let the students read the tale on their own, Sue asked, "There's a really interesting story here. Can anyone tell me what it is?"

A nun looked down and pulled the folds of her habit loose from some briars while several bashful children fidgeted, not expecting to be challenged to interpret nature. Aware that learning to read takes time and encouragement, Sue offered, "Only two creatures in Vermont, in New England for that matter, can create a scene like this. Anyone know what those two animals might be?"

A student, emboldened by this simpler question, responded, "The black bear. Bears can do this to a tree."

"Right, good, what else?" came Sue's immediate response, just seconds before another student blurted out that a moose, too, could bend trees down and strip them of the branches that grow eight feet from the ground.

Sue continued to let the students uncover the story by asking, "Now, why would a bear want to do this? Remember that bears always need to conserve energy and prepare for their long winter of sleep. Bears know that their reserves are limited, so to use the energy necessary to knock down a tree and claw at branches they would need a good reason. Can you think of a reason that would justify the use of all that energy?"

The students could think of nothing worth all of that effort for a bear. If it's not a bear, they figured, it must be a moose, the only other animal capable of this type of destruction. Chapter 1 of the tale closed.

The next clue to the mystery came from the fact

that all of the branches on the fallen tree pointed in the same direction as the trunk. The moose must have wrapped around the tree while taking it down. Sue continued to question the students, "A moose doesn't have hands, so how could it wrap around a tree and strip branches like this?"

The students, bolder now from past successes in deciphering this new language, quickly asserted, "The moose walked it down. It got the tree under its belly and walked over it." Sue's excitement grew with the students' as they became more attuned to the clues and more adept at eliminating possibilities. She turned to the standing tree and asked how the moose could have ripped off branches so high in the air. Surely, it didn't grab the trunk of the upright tree and climb up, stripping those branches with its underside, too.

"The horns, the horns, it did it with its horns." The students answered in chorus, several of them understanding at once and shouting out answers with growing excitement.

"So the moose is a male, right?" Sue questioned them. "Only male moose have a rack of horns large enough to do that. So let's see what the story says so far."

She ticked off the chapters of the story. "We have a moose, and it's a male because it has a big rack on its head. The moose was walking down a small tree, rubbing the tree with its underside, and all the while thrashing its head around. Because the moose did this on its own, let's assume it thrashed its head in pleasure. Now here's one last clue. This all happened in October, during the rutting season. During the rutting season, male moose look for mates; they're all worked up and ready to go. So why then? Why did this moose rub the little tree and thrash its head around with joy while doing it?"

The conclusion of the story began to show itself, first to the nuns who looked around knowingly and nervously, then to the children. The kids looked around as understanding lit up their embarrassed faces, none willing to share what they knew must be the only ending to the tale.

Matt, who had been silent all the time, the little tough guy who never offered anything in the classroom, seized his opportunity. He looked around at the other kids, his devilish streak flaring red inside, and said, "I know what happened. That moose," his words were anticipated by each nun's face, "that moose, he fucked the tree!"

Giggles rang out as the tension subsided throughout the group and Sue Morse smiled. Once again, she led a group of people to better understand, and to articulate, however coarsely, the workings of the nature in which they live.

Sue Morse takes people of all types into the wilderness to read the signs left by wildlife. An older guest of hers in the same situation as Matt offered, "The tree is a moose blow-up doll." Whichever way it's stated, it concludes the story of a fallen sapling and it begins the story of people becoming more intricately connected to their surroundings by learning how to communicate with them.

Sue runs Keeping Track. She is a self-trained forester, Shakespearean scholar, and animal tracker. She spent years in the woods of Vermont learning to use her eyes, ears, mind, and feet to move respectfully through the forest. She observes, interprets, and extrapolates the stories of nature in an attempt to learn how and where wildlife lives. Her interest lies not only in observation, but also in anticipation. She tries to identify the habitats most suitable for animal use by noticing which habitats are currently used by those animals.

She travels to the communities around her to teach people how to read their own surroundings. She believes that if people see with their own eyes how animals are woven into their communities, they will protect and share their homes with the wild creatures who also depend on the land for survival. Whether your neighborhood shelters moose or squirrels, rats or raccoons, it has a wild quality.

Not in My Backyard

For me, the wild quality of my community was the only thing worth saving. I'd be happy to see all of the boxy stores destroyed in the next quake. But for others it's the human character that needs to be preserved. John McCown has always held community as central to his life. He'll greet you with a crushing handshake and a beaming smile. If you look behind his smile, you can tell he's been through war—right here at home.

John's father worked alongside Martin Luther King Jr. in the early days of the civil rights movement. Eventually, he parted ways with Dr. King, believing that King should focus more on economic issues. His father felt that "it was great to be allowed to sit at the counter and eat dinner, but people needed to have enough money for the hamburger."

When he parted ways with Dr. King, John's father moved his family to Sparta, Georgia. The new home of his family became part of the first county in the nation since the time of Reconstruction to operate under African-American political control. Mr. McCown organized the town to elect the nation's first African-American probate judge, county commission chair, and superintendent of schools. "That's the type of background that I came from," John deservedly boasts. "My father died in 1976. I was sixteen at the time. After he died, the people in the town lost sight of the vision that he had."

Other community leaders adopted an apologetic attitude and accepted what came their way rather than seeking to build the community from within.

In 1983, John entered the military as an officer and set to the task of managing the handling of hazardous wastes by a crew of seven hundred soldiers.

"I had to organize it by compatibility and ship it off the base," he explains. "Even though I was dealing with hazardous waste, I didn't really know what it was. I didn't even care where it was going. It wasn't until I left the military that I learned that it was going to predominantly African-American communities like my own."

When he returned to Sparta, John found a community on the path to poisoning itself. Spartans considered plans for a $50 million incinerator, a nine-hundred-acre landfill that would collect ten thousand tons of garbage each day, and a one-thousand-bed prison. "This was their plan for bringing development to my hometown," he realized. "I was shocked. You're going to turn my father's legacy into a toxic waste site? I sure as hell wasn't going to let my town turn into a toxic waste site."

John called Lois Gibbs, director of the Citizen's Clearinghouse for Hazardous Waste. She supplied information about the effects of chemicals on communities with hazardous waste sites. "We went to every house and gave them info," John says, remembering his urge to awaken each person. "Three hundred people showed up at the meeting the next day.

"It took us five months to move the incinerator," John recalls. "Then we were targeted with the nine-hundred-acre landfill. It took us five years to defeat it. We did bake sales, quilt sales, fish fries, to pay $25,000 for our lawyer to try to halt it with the law."

Finally, the county of Hancock won against the

landfill in the Georgia Supreme Court. Four elderly ladies provided the convincing argument for the case, proving that the landfill would damage water and threaten their health.

After losing the case, the landfill developer approached the four ladies with $50,000 each to drop the case. John notes, "$50,000 is a lot of money to those ladies. But they said no. Then he offered the attorney $75,000, and remember, the town owed the attorney $15,000 at the time. The attorney said no." The developer upped his offer to the ladies, suggesting they take $75,000 each. Again, they stood firm in their decisions. A second shot at the attorney, this time with $100,000, also proved fruitless.

John acknowledges, "There is no landfill in Sparta because of those four old ladies who would just not be bought."

After fighting to keep environmental foes out of his community, John faced foes within the community. "The local power structure turned on me. They black-balled me. I couldn't find work in a five-county radius," he says with a degree of amazement. Even after his military career, he couldn't find work as a corrections officer because prison officials looked at him and said, "There's no way we're going to hire you. You might organize the inmates."

John survived for a few years by selling used fur-niture at flea markets across the southeast. He bought truckloads of sofas at Salvation Army auc-tions, learned upholstery skills, and sold the rehabilitated sofas to low-income communities. John still believed in his community despite the treatment he received at the hands of its leaders.

"I dealt with it for five years," he says of the situation. "I was about to reach the breaking point, so I caught a ride to Atlanta and I applied for a job with the Sierra Club."

Two weeks after he applied, the Birmingham, Alabama, office of the Sierra Club hired him as an organizer of struggling communities in the southeast. He was chosen to establish links between two groups that have had little contact in the past—a national, primarily white organization and the local, primarily black communities of the South. He says, "They have a lot in common. I am trying to pool these resources and make a coordinated effort to work together to take on our common foes. I'm a peacemaker.

"I'm most proud that I've been able to lend badly needed assistance to suffering communities—basic organizing experience, grassroots organizing, media relations, proposal writing, you name it. Pollution does not respect the barriers that keep Americans divided among themselves. The pollution will eventually migrate into white suburbs and poison white babies, too. If we don't come together with communities now, and work together to protect our common destiny, there won't be any suburbs far enough out."

The struggle to protect community is a battle for

quality of life. If the environmental movement hopes to succeed in the twenty-first century, it must build a movement of neighbors, from every neighborhood, one neighborhood at a time. Sue Morse, John McCown, and the teachings of Gandhi lead the way. Will we follow?

Gandhi's Got It Going On

Nonprotectionist, antisprawl, radical localism will explode as we head into the next millenium. If we understand ourselves well enough to express our needs to large corporations they will re-engineer their strategies, placing local products on their shelves next to national brands and listening to the concerns of a citizenry interested in keeping its communities centralized. The Wal-Marts of the world will demonstrate their responsibility to local communities if we hold them accountable for catering to our needs.

As I traveled throughout the western part of our country during the 1996 presidential campaign, I was struck by the amount of hate directed at Californians. Campaign literature for environmentally conscious candidates was marked by stinging epithets that urged people, "Don't let Portland, Oregon, get Californicated!"

I grew up in ground zero for suburban sprawl in my small town on the outskirts of Los Angeles. I've lost my home already. This is your chance to protect yourself and the community from which you draw both satisfaction and pride. Go radically local. Get to the root of the problems that threaten our communities and our environment.

Someone once asked Gandhi, "What do you think of Western civilization?"

Wrapped in a dhoti, his bones poking out the tight fabric of his skin like tent poles, he paused a moment in thought. "It would be a good idea."

Pop Quiz

Pop quiz. You're on a big yellow school bus. It's the kind with seats that barely have enough foam on them to cover the metal bars underneath. On top of the thin foam is the green plastic that never heats up in the winter and in the summer sticks like a snail to any exposed flesh that might touch it. Gum sticks to the undersides of everything, and scratched into the backs of many of the seats are scrawlings like "JP's MOtheR is a ToAD" and "JoHNniE liKes GiRRrls."

It is a hot spring day, the first when you believe that it might not snow anymore and begin to think about where your swimsuit was stashed for the winter. The sun beats through the greasy handprints on the windows. Your forehead beads up with little droplets of sweat. You can't get your window down because the buttons that should slide in to release the window are stuck. A few seats in front of you, someone has managed to open his half window, and fresh air pokes its way tentatively into the bus through the crack. Up front the bus driver eyes the crowd through a gigantic rearview mirror and communicates silently with pointed fingers and crunched up facial expressions.

Andy Kimmelman is sitting next to you, unwrapping a molten candy bar. The melted chocolate of his Charleston Chew strips itself off the bar as he uncovers it, and he licks the plastic to recover every bit of mushy sweetness that he can. He sees you looking at him, mistaking your look of disgust for one of envy, and offers you a lick.

The bus travels quickly down a winding road, bounded by a sheer cliff. Mrs. Rutberg, the bus driver, anxious to end her day and return to her own children, grinds the gears down with a scraping growl as she slows around the turns of the road. And yet the bus picks up speed and hurls itself through a ride you have passively taken since the first day of school in September.

Then, all of a sudden, you feel tension in the school bus. A little girl begins crying and a boy in overalls points to the front of the bus. You watch the bus driver weave in the front seat and fall from her post at the wheel.

The bus picks up speed as it rolls along the downside of a slope, its brakes and gearshift unattended. It drives straight toward an S-curve and its wheels kick up gravel from the side of the road like corn kernels in the chute of a hot air popper. The bus weaves wildly, testing the extremes of the road and unable to return itself to a steady, consistent pattern of progression.

You look around and see the kids ask themselves, "Why is this happening? This has never happened before. Where's Keanu Reeves when you need him?"

Some kids think about jumping out a window. Others are locked on this catastrophe like a deer caught in the headlights of an oncoming car. Some are simply dumbstruck.

Everyone else, like you, is just a kid. You can't vote. You can't drink. You don't know how to drive. You don't have a license. You're in junior high, after all. Someone else has always been there to take care of things. Someone else has always driven the bus and made sure that you got home safely to your home, your baseball glove, your Barbie, your rope swing down at the river.

What do you do?

Do you watch in disbelief as the bus caroms off of a tree, ripping branches from the trunk? Do you hope that the bus driver will pop back into her chair with a laugh? Do you scream for help? Do you wait for someone else to take care of the situation? Or do you jump over the kid next to you and hop into the driver's seat?

Someone has to save the bus. It's picking up speed and before long it will be beyond saving. You know that you need to save it to save yourself and everyone who shares the roller coaster ride with you. You can't be sure that anyone else will jump behind the wheel and

grab the gears. And even though you have never driven before, never mind driven a big yellow school bus,
neither has anyone else around you. Your only chance
for survival rests in your own inexperienced hands.

The world needs bus drivers. It's veering off the
road and there's no one at the wheel.

Pop quiz. What do you do?

SECTION ◇ 9

Straight Up,
No Distractions

*T*he greatness of men and women is shown in the actions of those whom they inspire. David Brower is more than the most successful environmentalist who ever lived; he has inspired generations of environmental leaders. If not for David Brower, the Grand Canyon would resemble a bathtub—flooded by an ill-conceived dam that seemed inevitable before he weighed in. If not for David Brower, our world would be less hospitable, less magnificent, less magical.

David is a radioactive core—when focused, his energy is unstoppable; when unfocused, it consumes anything and everything. David managed my campaign for the job of president, and his support moved the tides in my favor. After having been fired as the Club's executive director in 1969, David had finally regained his standing within the Club. One week after I began my job, David called a reporter for the *San Francisco Examiner* and announced that he was angrily resigning from the board of directors. The media stormed our offices, asking what I had done to make him resign. I had no idea that he was even upset. I drove up to his house in Berkeley to beg him to stay on. He did, saying that he wanted to give me a chance to fix what was broken in the

Sierra Club. It made for an exciting first week on the job.

I first encountered David in my junior year of high school, through John McPhee's *Encounters with the Archdruid,* a book that pitted David against three of his natural enemies—a dam builder, a developer, and a miner. David was humanized and heroicized. Although McPhee revealed that David lived in a house made out of ancient redwood trees, he also showed Brower morally debating and trouncing those who would destroy the environment. I wrote David a letter with the same hope as a child who writes a letter to Santa Claus—never expecting a response, but feeling good for having done so. A week later I received a typewritten letter from David himself. The first sentence scolded me for having placed a comma before a parenthesis in my letter. He then told me to remember Walt Whitman's admonition, "Resist Much, Obey Little."

When David speaks to you his eyes hold you firmly in the crosshairs as his words shoot straight to their mark. For a rabble-rouser, his voice is quiet, patient, and even. He speaks with the measured meter of a sage, he delivers timely punch lines, and he moves on to another point as soon as understanding lights in the face of his audience. Hearing him speak is like listening to a tribal historian. He chooses stories from his enormous repertoire of experience and weaves them into a broader view of life.

After all of his personal achievements, he's still

fond of Tom Hayden's line, "All I have been able to do in my life is slow the rate at which things get worse."

On any given day, you may find David eating lunch at Sinbad's, a quiet restaurant at the foot of Mission Street with royal blue booths that preside over the San Francisco Bay. He holds court, sipping a Tanqueray martini (ordered with "no distractions"), as young environmentalists bring him news of issues and controversies. He offers his expertise, often referring them to other appropriate leaders for help.

Born in 1912, David is well aware of his age. He is somewhat uncomfortable at having become part of the old guard with whom he has always sparred in an attempt to broaden minds. But he adheres to Ansel Adams's saying, "If you're going to get old, get as old as you can get." He intends to be with us, working for a healthy planet, for a long time.

Werbach: Okay, let's start simple. What's your name?

Brower: No comment.

Werbach: No comment? Wow, Dave, I thought that after all the interviews you've done you could at least answer that.

Brower: [in deadpan monotone] Good morning, Adam.

Werbach: I want to get across a few things during this brief time. First is some of the lessons that you've passed on. Second is your prognostications for the environ-

mental movement, where we're heading, where it's been. And third is talking
about solutions and hope. In the mean-
time, I plan to challenge you on everything that you say.

Brower: Let me start with my favorite war story
because it provides a good analogy.
World War II was over–we were driving
out to a pass between Italy and Austria.
It turned dark and we were in a regi-
mental combat team, four miles of vehicles, and we turned on our blackout
lights so that the enemy couldn't spot
us. They were essentially useless to us. I
got on the radio and asked the regi-
ment's commander, "Why the hell are
we driving blackout? The war is over."

Colonel John Hat responded, "Dave
you've got a point." So he gave the order
to turn on the lights. So, all the lights for
this four miles of vehicles went on. We
were nearing the pass in a wide open
space where there was no shelter, just a
big, long meadow winding several
miles. We later found out from our
scouts who were there ahead that the
Germans at the pass who did not know
the war was over had their artilleries
lined on us. They were going to fire and
then our lights went on, and they
figured we knew something they didn't
know.

So the lights had saved my life and a
lot of other lives because I simply asked
if the lights could be turned on.

The moral of this is that right now in
our society, we need to turn the lights

on. We need to do it in a hurry because we're not seeing our way.

If things go on as they're going now we will need to produce as much food in the world within the next forty years as was produced in the last 8,000 years. No way. That's what mindless growth has done to us. There are a few lights being turned on now. We need to get many more lights on, but a lot of people think all the old games are still the same games to play.

Werbach: What are the old games?

Brower: The old games are just old. We worry about clean water, clean air, wildlife. But we don't worry about what's happening to society as a whole, in all the cultures and all parts of the world where we are overstraining the Earth in a way it was never overstrained before. We're going on with the assumption that we can go right ahead doing what we're doing. We absolutely cannot. We need to take a fresh breath and look again. It is a different world from what we last looked at. It's changing very rapidly.

Werbach: What's your definition of an environmentalist? What is a new environmentalist?

Brower: I still don't like the word.

Werbach: Neither do I. What do you like better?

Brower: I just like being a conservationist. That's what it was until Earth Day. And then people wanted to add another syllable to it.

Werbach: You're referring to Earth Day 1970. That was the first push to connect pollution to the conservation movement. And that changed *conservationist* to *environmentalist.*

Brower: Well, that was just Earth Day 1970. Eight years before that Rachel Carson had written forcefully about pollution from a chemist's point of view.

Werbach: Did she call herself a conservationist?

Brower: No, she was a biologist. Conservationists didn't exist until the Theodore Roosevelt administration, when the name came into use. It may have been used earlier. We have trouble with names. Rename something, give it a label, and it loses some value. I say don't think about the label, but about the true purpose. So, you ask me what is an environmentalist. I'm stalling. I'm not answering your question because I'm not quite sure.

Werbach: Well let's talk about what environmentalists, conservationists . . . whatever . . . think. Let's think about a kid who lives in the city. She grows up in South Central Los Angeles. She certainly hasn't heard of the Sierra Club or David Brower. She's dealing with a tough life of poverty. Should she care about the environment? Should the environmental movement care about her?

Brower: I spent years trying to save the California condor. And an African-American woman said, "Saving the condor isn't going to feed my children." I should have said that killing the condor isn't going to feed your children either, but I was years late coming up with that answer.

Werbach: Tom Hayden has said that all that he has been able to do during his career is slow the rate at which things get worse. What have you been able to do during your career?

Brower: That's all I've done. That's all the Sierra Club has done. That's all the environmental movement has done, along with all other movements.

Werbach: Slow the rate at which things get worse? That's not a very hopeful thing to say. I mean, here I am, I'm twenty-four, and should I look forward to sixty years of only slowing the rate at which things get worse?

Brower: No. No, absolutely. The thing you have to do is switch that. That's not enough. We've got to increase the rate at which things get better. We can do that. But we can't do it by following old systems. Because the old systems are what make it worse. They refuse to give a value to things we can't replace. That's why I advocate global CPR—conservation, preservation, restoration.

Werbach: CPR and Green Plans, like the Dutch have done?

Brower: Yes. We need to consider our conservation goals a little bit more carefully and make sure they are more comprehensive. We need to preserve things we can't replace. And we need to restore. It never occurred to me that we could put things back together.

Wendell Berry has pointed out that it's very arrogant to assume that you can restore a natural system. That's about the only place I disagree with him. It is possible. You can do it, and it's being done. We need to describe where this is being done, where it's working, and where we haven't made worse mistakes by trying to correct the earlier mistakes.

This is a possibility. This is a human possibility. Your generation has got to get going on it. And the Sierra Club has got to get off its duff on these things. It hasn't. It put the word *restoration* in its own charter recently. But it hasn't done anything about it.

Werbach: Well, I think it took a step at your behest when it advocated the draining of Lake Powell. But getting back to my question, what is the role of environmentalist?

Brower: To take a broader look. For example, think about the value of a tree. What's a tree worth? The marketplace will tell you what a tree is worth for pulp or for two-by-fours. That's all the marketplace tells you. Nothing about the value of the forest and the balance of carbon dioxide on earth—one of our big problems right now. We don't know how much oxygen we can do without before we're in deep trouble. We haven't any numbers on that. It doesn't tell you anything about the value of water quality and quantity that forests provide. It doesn't tell you anything about keeping soil in place, and soil is rather important. It doesn't tell you anything about habitat. The forest is habitat for millions of species, most of which we haven't discovered yet, but that doesn't mean that they are not there. And the marketplace doesn't give you any value for beauty. So the marketplace counts pulp and two-by-fours, but none of these other vital resources.

Werbach: Well, we're talking about comprehensive, big-picture thinking. Certainly things that you weren't talking about, I'd say, in the same level of detail thirty to forty years ago. In some ways, people accuse comprehensive thinking of not being as inspirational as talking about a place. As talking about saving the Grand Canyon from becoming a water tank.

Brower: I began to get more general thinking, I guess, in the battle to save Dinosaur National Monument. I learned it from Howard Zahneiser, who was the executive secretary of the Wilderness Society and the principal architect of the National Wilderness System. He said, "If we're going to save wilderness, we've got to use nonwilderness areas better." That's the story. You can't replace what's in wilderness. You don't have to explore it for other uses. You can use wilderness as wilderness. Let's just let it be wilderness. Pretend it doesn't exist for other purposes. We'd get by if we didn't have the state of California; we'd somehow get by without California.

Werbach: What is wilderness good for?

Brower: It is the ultimate encyclopedia. It is the source of every living thing we have. Every ability that we have came from wilderness and we shouldn't forget that and get rid of the source.

And I am not giving away any secret to tell you that you are the product of the union of one lone egg and a lucky sperm. Those two bits of genetic material knew everything that was necessary to build you. A hundred million rods and cones in each retina, all put in the right place, letting you behold creation and do something about it. Your eye can encompass the whole damn world, big chunks of it all at once. You see it all, get a good grasp of it, you are pretty comfortable with taking it all for granted.

You have all of this ability, and the ability to remember faces of people you know in your own private video in your head. And when you are eighty-four, you have got a hell of a lot of faces that you can conjure up. Or old songs, complete with lyrics and music. And you have the ability to know and use compassion and love.

That is, all that stuff is stored. How is it done? That lucky sperm and that egg know all about it. No guesswork. There it is. It works. You say it is highly unlikely but it has happened a good many times. For a hundred billion people it has worked well so far.

Werbach: It's always been there. But David, you have been called an archdruid. Aren't you someone who cares more about trees than people?

Brower: A developer who wanted to develop the hell out of Cumberland Island called me a druid. John McPhee added the "arch."

Werbach: Do you care more about trees than people?

Brower: I love the line I got from Charlie Callison, a vice president of Audubon. Do you like birds more than people? He said, "I like people who like birds." Well, I like people who like trees. People who don't care about trees? I don't like them so well.

Werbach: How do you institutionalize impatience? In my mind that is one of your greatest qualities. You're terribly impatient.

Brower: I was, and I'm more impatient now at eighty-four. My passport doesn't expire until the year 2004, and I intend to get it renewed. We're not really sure I'll have a chance to renew it. I want to see something happen before I check out. How much is impatience related to caring? I think caring is a good word. People have to care. Compassion can happen next.

Werbach: Well, you're coming upon the recipe for inspiration. And one of the signs that I think you're developing more and more now is talking about comprehensive solutions that aren't just fighting and stopping that which is going to get worse. You're talking about solutions that address the whole number of things that work toward that. You started by taking people out into the Sierra to see the wilderness.

Brower: I did that. But that was the idea of Muir and Colby. It wasn't my idea. I just was one of the Johnny-come-latelies in taking people out into the wilderness. I took some four thousand people out into the wilderness and back—except for two who died. But the whole idea is just to point out: You're here, you've seen what it is like. Let's save it and celebrate it.

Werbach: So, step one, know the land. Get out and see it.

Brower: So you get to love it, learn what the dangers are and what can be done about them. Those three steps are essential. Those are the three essential steps in the books that we published back when I was doing it. Get people to love it there. Then let them know what's happening in that same book, and then let them know what they can do about it.

Werbach: Well, that to me is the brilliance of the Books Program and what we need to innovate on right now. You took an idea that was abstract to most people, wilderness. Most people don't get a chance to go out into the Sierra and bond with the wilderness. With your books they could see why it was important. They could look at the pictures and understand it. Then you follow up quickly with what they could do about it. What about right now? If you're try-

	ing to reach someone who's sitting in her living room in Sioux City, what's the best way to do that?
Brower:	Sioux City?
Werbach:	Yeah. And you want to get them involved in the environmental movement. Why do you try to reach out to them? Why should you reach out to them? Should you reach out to them? And how do you do it?
Brower:	Well, in Sioux City, I would ask Don Pierce to figure out what you do in this flyover country. He knows how to handle it. He lives in it.
Werbach:	So, you start with somebody who knows the land.
Brower:	Someone who knows what the thinking is in that area. I know something about the thinking on the coast because I grew up along the coast. And I'm not comfortable unless I can look at Mount Tamalpais.
Werbach:	Organizing needs to be locally based?
Brower:	It helps. But this is the other thing that you must remember: We would have no national parks if we had allowed local interest to govern. Every national park we wanted had local opposition. It took national interest to get them. This morning's paper has the same story again. Local interests in San Luis Obispo want to put in a resort development near San Simeon along the Big Sur coast. That is a global treasure, it's

incredible. The local interests want to get quick money out of the resort. We haven't got that adequately protected yet. The Forest Service is not able to protect it adequately. The Yosemite Park idea was opposed in Mariposa. And I was just down in Arizona, Sedona—a beautiful place, Oak Creek Canyon. It should have been a national park. It has been absolutely clobbered by developers.

Werbach: What's your criterion for a national park?

Brower: There has to be something uniquely and locally significant in the natural beauty of the place. Of course, something really spectacular is outstanding. It needs to be outstanding. When I thought of that originally, I didn't think that a prairie could be outstanding. I think so now. I've learned a great deal more about what goes on in a prairie than I knew. Today I would just as soon see the Earth National Park I advocated in the big *New York Times* ad the Club ran in 1964. We should look at the whole Earth as if we wanted it to remain as beautiful as it is.

Werbach: I am becoming more and more convinced that the future of the environmental movement will be increasingly local. Our role will be helping to reestablish a sense of community that people have lost. That's the sense of radical localist. Is this something that you endorse?

Brower:	It seems that that word *community* is the word that we need again. It's a good word. We'll probably wear it out soon. But community is what we need now. And this is one of my worries with what we're getting into now as we get less and less community, as we destroy local habitat and join the stampede to globalize trade.
Werbach:	What do you do to stay sane?
Brower:	Tanqueray martinis.
Werbach:	Straight up, no distractions.
Brower:	I am incredibly fortunate. I have been fortunate to have been married for fifty-four years to a very patient woman. That keeps me sane. That doesn't necessarily keep her sane.
Werbach:	What advice do you have for the leaders of the future?
Brower:	Careful. I'll give you some bad examples.
Werbach:	The first bit of advice you gave me was, "Don't take my advice." Or, at least, "Don't take too much of my advice."
Brower:	And I certainly gave you Rule Six.
Werbach:	What is Rule Six?
Brower:	Never take yourself too seriously.
Werbach:	Yes.
Brower:	And you ask what are the other rules?
Werbach:	What are they?
Brower:	There are no other rules. But I am glad

you make it a rule to travel around as much as you do.

Werbach: What are other rules that you live by?

Brower: Well, I still like the old business that if you get power, pass it around. The only problem with that is that it gives you still more power. But just don't abuse it. Because it does corrupt. Corrupt somebody else with it.

Werbach: And what else?

Brower: I'll give you one of my favorite quotations. Every time I close a talk with it, I get a standing ovation. If I forget, I don't. It was in Adlai Stevenson's last speech, July 1965, when he was our ambassador to the United Nations:

We travel together, passengers on this little spaceship, dependent upon its vulnerable reserves of air and soil, committed for our safety to its security and peace, preserved from annihilation only by the care, the work and I will say the love we give our fragile craft. We cannot maintain it half fortunate, half miserable, half competent, half despairing, half slaves of the ancient enemies of mankind and half free in a liberation of resources undreamed of until this day. No craft, no crew, can travel safely with such vast contradictions. On their resolution depends the survival of us all."

CONCLUSION

Hish-shuck-ish-tsa walk.
"Everything is one."

"Everything is one."

*T*ensions were mounting in British Columbia. I had been hearing rumors that the local Sierra Club chapter wasn't doing enough to protect the last intact temperate rain forest in the world while thousands of activists from around the globe promoted the cause. The habitat of grizzly bears, sockeye salmon, and the elusive white "Spirit Bear" was threatened.

Vicky Husband, a Herculean volunteer who never sleeps and always seems to be in a constant state of motion, greeted me as we gathered for a fact-finding trip to reorganize our efforts.

"I'm taking you to meet the Nuu-Chah-Nuulth tribes first. You can hear what they have to say," she said.

The British Columbia rain forest isn't just the home of exotic threatened animal species. It is home to the First Nations, the native tribal bands who trace their presence in the forest back ten thousand years. To this point, many environmental activists had failed to consider them. Bobby Kennedy, from the Natural Resources Defense Council, is one environmentalist who has worked to build bridges with the First Nations. Still, as fast as he built the bridges, others burned them behind him.

In their years of living in the rain forest, the First Nations never faced challenges as great as the challenges of today. Eighty percent unemployment, staggering alcoholism rates, and constant battles with the Canadian government about treaty rights have nearly extinguished the hope of the community. The loss of salmon streams and their right to fish and hunt as they have for thousands of years seeds a bitter resentment in the people. They see no opportunity, and in the band of the Heiltsuk, comprising of fifteen hundred people, a young person commits suicide every month. After ten thousand years of surviving in the cradle of the rain forest, the First Nations now barely hang on.

The most extraordinary characteristic of the British Canadian temperate rain forest is its size. The temperate rain forest covers seven million hectares around the coast of British Columbia, the size of Washington, Oregon, and California combined. We flew for hours over the greatest stretches of intact wilderness that I had ever seen. Remoteness has protected the land in the past, but distances are shrinking in this age and timber companies like Western Forest Products and Interfor sharpen their saws and prepare to enter the remaining intact watersheds. Grizzly bears require hundreds of miles of habitat to survive. As the forest splinters into fragmented islands separated by logging roads and clear-cuts, the grizzly's future becomes bleak. But the First Nations have needs as well.

Our first meeting was with the leaders and chiefs of the Central Region tribes of the Nuu-Chah-Nuulth, the thirteen tribes that surround Clayaquot Sound and Northern Barkley Sound. The meeting was held in the small town of Tofino at a hotel, the Tin-Wis, owned by the Tla-o-qui-aht. I expected a rustic lodge. I was surprised to find that the Tin-Wis was operated by Best Western, Inc., and had all imaginable amenities. As we sat around a conference-room table, snacking on egg salad sandwiches on white bread, we began the conversation. Tension hung in the air.

The tribal bands had been at the center of the British Columbia forest controversy over the last thirteen years. When logging companies began entering their territories, dedicated European environmental activists blockaded the loggers' progress.

Cliff Atleo, a member of the Ahousaht, looked at us with the stern gaze of a man who had been disappointed many times. "We thank you for coming to meet with us. While we appreciate the work you have done, we feel we have been ignored by the environmental organizations."

The top priority for the bands was pulling themselves out of the economic sinkhole in which they were stuck. Economic development was needed, and the timber companies were the only ones offering it.

The bands were angry that Greenpeace had broken protocol by not communicating with the tribes about their plans. The issue to them was not one of trees, but one of sovereignty and a right to survive as a people.

Cliff Atleo asked me to explain the history of the Sierra Club. I explained how we had started with the designation of Yosemite as a national park.

"I don't believe in parks," Cliff said sternly. He paused to let me consider his bold statement. I was beginning to get nervous. If the bands supported the logging, there was little hope of protecting the rain forests. The pause seemed to last for minutes, as the tribal leaders stared us down.

"If you start a park," Cliff continued, "you are saying that it does not matter how the land is managed outside of the park. We believe in protecting all of our land."

The tone of the conversation began to shift. The leaders explained their challenges and asked whether the Sierra Club could help them. Steve Charleson of the Hesquiaht tribe asked about the founding precept of the Sierra Club. I quoted John Muir's idea, "When you pick anything up by itself, you find it hitched to everything in the universe."

Steve nodded. "We too have a saying: *Hish-shuck-ish-tsa walk,* which translates to 'Everything is one.'" Steve explained how three of their salmon streams had been choked to death by avalanches, the result of clear-cutting on the steep slopes of

Hesquiaht Harbor. The timber company and the government promised to help restore the streams, but there had been no help forthcoming. Vicky quickly offered to apply pressure to the government in Victoria to incite them to action.

Another tribal leader explained the destruction of his family's traditional fishing grounds. Giant trawlers from the East Coast swept the bottom of the ocean floor with their nets, destroying the reef that once protected a school of ling cod. For generations, his family enjoyed a bounty of cod so rich that each time they cast a hook they would catch a fish. Today the coral has been leveled by the "draggers" and there are no more fish.

The meetings ended on a positive note. The bands made clear that they were to be consulted and respected before any action was taken in their territory, and they asked us to remember the economic hardships they felt. Before we left they advised us to tread carefully as we headed north. "Your reception will not be so warm there," Cliff warned.

We boarded our plane and headed north over Flores Island, the next potential target for logging. The island was pristine, resembling the TV paradise of Fantasy Island. Whales spouted off the coast as if to welcome us. As we continued north we flew over the Hesquiaht territory and saw the damage that Steve Charleson lamented. The land looked like it had been torn apart by a tornado. Lush green forests were converted into collapsing muddy slopes. Mudslides deluged the delicate river valleys. The land was pillaged.

We flew into the small village of Bella Bella to refuel and meet with the band council. It was late, so we went straight to the home where we were staying. That night we met with Frank Brown, a young Heiltsuk man who was determined to get beyond the rhetoric that had surrounded the issue of saving the forests. He was frustrated with environmentalists who thought only of the land and not of the people. He was frustrated with timber companies for the devastation they would leave in their wake. Bella Bella had just entered the commercial timber age. The timber companies, having exhausted the easily accessible supplies near Clayoquot, headed north. Frank Brown refused to wait for someone else to solve his problems. He had started a tour company to help celebrate Heiltsuk culture. He was training eight young men from other tribes how to do the same so that outsiders could see the thriving cultures the tribal bands created in these remote regions.

"You environmentalists forget," he said. "We'll still be here when it rains."

The next morning the clouds of the rain forest hung low over the town as we walked toward the town meeting hall to meet with the band council. We had been warned not to wear anything identifying us as environmentalists. An old woman stopped as we passed and asked, "Have you come here to catch our sickness?" A man on a bicycle passed by and said, "The land is not for sale."

We arrived precisely on time for our meeting, not wanting to do anything to offend. We were purposely made to wait for half an hour before being ushered in to meet Arlene Wilson, the chief counselor. Arlene seemed somber and angry. We took our seats quickly.

"I want to know what your agenda is," she began. "I want to know what you want from us." The other leaders nodded their heads in agreement. "You environmentalists claim to have high principles, but it is we who have to put them into real life." She looked out the window. Two eagles floated across the breeze. "I do not trust environmentalists. Last year Greenpeace and Rainforest Action Network said they were coming here to help us. Then they broke protocol. They didn't communicate with us. They showed us no respect. Our band council voted not to have environmentalists in our territory. I supported the motion to deny Greenpeace water and gasoline if they came through."

Our group collectively swallowed hard. If the Heiltsuk decided to ban us from their territory, our trip was finished. Without gasoline our plane would be grounded.

Arlene continued. "And others have stronger feelings." She nodded at Ed Newman, the band's chief treaty negotiator. His face was a rock of determination and anger.

Ed paused a long time before saying a word. He stared directly at me, never blinking, ignoring everyone else in the room. He finally spoke, slowly, deliberately, with the force of a glacier. "I have no use for environmentalists," he said. "If you want to help, I say stay the hell out of our territory. You say you want to help us with tourism. We don't want your carpetbaggers—your German tourists who foul our beaches and take our fish. We have seen you environmentalists fight against logging. When you fight against logging, you fight against us. It is our turn for economic development. We have suffered miserably for too long. You try to deny us a livelihood. You take food from our children's mouths. You're trying to stop us from improving our condition. You are no different than those who tried to kill us with their guns. It is our right to assert our control over our traditional territory. You environmentalists have no place challenging this right." He crossed his arms, never taking his eyes off me.

He continued. "I have heard that you met with an individual last night. By doing so, you broke protocol. You offend us by coming into our territory to meet with us and ask us permission to travel while meeting with our people behind our backs. You cannot come in here and tell us what to do."

Ed leaned back in his chair. "Now I have given you something to think about."

His words hung in the air.

Arlene finished the meeting. "Now we have other work to do. Because we are a compassionate people we will sell you gas and food and allow you to travel."

Shell-shocked, we trudged back to the plane

silently. I felt like I had been punched in the gut. Our intentions were never to hurt anyone, but in our rush to do the right thing we had trampled people. Larry Jorgenson, a member of the Heiltsuk whom we met later, explained that unless we were willing to commit ourselves to working with their community, we would be unsuccessful in protecting the forests. "Your challenge," he said, "is to understand *Go'alah*. It is the art of seeing everything, while focussing on nothing in particular."

I learned an important lesson on the trip. Environmentalists, in a rush to save the natural world, often forget to consider the implications of their actions. Our dedication leads to a self-righteousness that bowls over friends as well as enemies. I learned to respect, once again, the local perspectives of our activists on the ground. Vicky Husband's advice had been sound. I needed to hear the anger myself to understand.

Unless we're willing to listen, we may be as destructive as those who would destroy the earth. It's time for us to work together.

In order to create a world that we are proud to leave for our children, we need to build bridges between people of goodwill that are supported by our common urge to protect the environment. We need to get ahead of the curve of destruction. It's time to evolve from an aging movement that reacts to emergencies to a youthful movement with "activision." Activision is activism powered by a broad vision for future generations. Activision follows the principle of *Go'alah,* sometimes forsaking the details for the big picture.

There are new ways to find permanent solutions to our environmental challenges. First, the public needs to demand their basic environmental rights. Second, the public must force politicians to earn their paychecks by creating innovative solutions to meet those demands. The Dutch have already proved that forward-reaching solutions are possible.

In 1990, the seal population in the Netherlands began dying off in record numbers. The Dutch people immediately demanded action from their government. In response, the Dutch government commissioned the RIVM, the national scientific agency, to study trends in the natural world and recommend actions to sustain the environment.

After just one year, the RIVM released a report that called for a pollution reduction of 80 to 90 percent.

Because the RIVM was well respected, industry could no longer make scientific excuses for their own lack of courage. The RIVM eliminates the question of "if" from the equation. Once this nonpartisan scientific group establishes that a problem exists, the country wastes no time in long debates over whether or not certain actions threaten the environment. Representatives don't eat outlawed chemicals in an attempt to prove their harmlessness. The bickering over whether or not water is hazardous or air is dangerously polluted ceases.

Instead of shying away from this radical goal, the Dutch designed the National Environmental Policy Plan (NEPP) to make the Dutch environment sustainable within one generation. By tackling waste issues, toxics, air and water contamination, recycling, and the global threats of ozone depletion and climate change now, the Dutch hope to provide the next generation with a basis for continued sustainability.

In return for a binding agreement to achieve goals of radically reduced pollution output, each industry was given the freedom to pursue these goals by its own design. No technology restrictions were applied. No definition of the path to the end goal was set forth. The allotted time of twenty-five years allowed companies to set short-, medium-, and long-term goals and to pool their resources to invest in new technologies to replace the environment-threatening old.

When a company falls off the path to sustainability, the RIVM may exact heavy fines and outline a more structured, inflexible, and rigorous path. Environmental groups receive funding from the government to act as watchdogs of the operation.

A five-year review of the NEPP shows a pollution trend toward sustainability. The rest of the world is capable of achieving the same type of radically comprehensive, solution-based environmental plan. But first the environmental movement must unite from within. If environmentalists cannot come to an agreement as to which goals to pursue, businesses cannot be expected to follow their lead. Instead of arguing about the science, we can focus on the solutions.

Fortunately, a new breed of environmentalism is rising. The world that spawned the old movement has changed. Jerry Garcia is dead. Geronimo Pratt has been freed. When you talk about McCarthyism, most people under forty think you're talking about Jenny McCarthy's TV show. There is no need to force the language of revolution on a public that largely agrees with the agenda. We can use religion, pop culture, business, and politics to bring people together to protect the planet. We have an opportunity to bridge the gaps in our society—to help the Heiltsuk band achieve economic development, to help Reverend Conley achieve justice for his community, to help Clara run a mile, to help every person believe that he or she can affect the future.

This book is only an outline for the steps we need to take as a society. You can create better models than those used by the people featured in this book. As the old saying goes, a journey of a thousand miles begins with a single step. When you put this book down, take a walk around your neighborhood and introduce yourself to someone you haven't met. Ask the manager of the local grocery store to stock local organic produce. Make a call to the office of your local member of Congress and thank her for serving the community. Plan a weekend camping trip. Every time you stand up for something, someone will try to bring you down. Take my advice, when you're doing the right thing, sometimes it's better to ask for forgiveness than permission.

Don't forget to take care of yourself. If you neglect to get out and enjoy this beautiful world you will lose your capacity to be a clear voice to protect it. You don't need to become a rabid activist. I've learned it the hard way, by losing good friends to this job. I've spent too many all-nighters working on this book, too many days away from birthday parties and family reunions, too little time just hanging out with my friends.

You have to save yourself before you try to save the planet. You've got to have fun and be a little wild.

Henry David Thoreau's most frequently misquoted saying is, "In wildness is the preservation of the world." Often, the word "wilderness" is accidentally substituted for *wildness*. But the word *wildness* is crucial to his meaning. Wildness encompasses wilderness—canyons and forests and waterfalls and mountains—but it also concerns something wild within each one of us. This wildness allows us to make mistakes and to try something new. Our wildness ensures that at the moment when the future seems hopeless, we will find a solution. The time has come for wildness.